Electric Vehicle Technology Explained

Electric Vehicle Technology Explained

James Larminie
Oxford Brookes University, Oxford, UK

John Lowry
Acenti Designs Ltd., UK

John Wiley & Sons, Ltd

Copyright © 2003   John Wiley & Sons Ltd, The Atrium, Southern Gate, Chichester,
West Sussex PO19 8SQ, England

Telephone (+44) 1243 779777

Email (for orders and customer service enquiries): cs-books@wiley.co.uk
Visit our Home Page on www.wileyeurope.com or www.wiley.com

All Rights Reserved. No part of this publication may be reproduced, stored in a retrieval system or transmitted in any form or by any means, electronic, mechanical, photocopying, recording, scanning or otherwise, except under the terms of the Copyright, Designs and Patents Act 1988 or under the terms of a licence issued by the Copyright Licensing Agency Ltd, 90 Tottenham Court Road, London W1T 4LP, UK, without the permission in writing of the Publisher. Requests to the Publisher should be addressed to the Permissions Department, John Wiley & Sons Ltd, The Atrium, Southern Gate, Chichester, West Sussex PO19 8SQ, England, or emailed to permreq@wiley.co.uk, or faxed to (+44) 1243 770620.

This publication is designed to provide accurate and authoritative information in regard to the subject matter covered. It is sold on the understanding that the Publisher is not engaged in rendering professional services. If professional advice or other expert assistance is required, the services of a competent professional should be sought.

Other Wiley Editorial Offices

John Wiley & Sons Inc., 111 River Street, Hoboken, NJ 07030, USA

Jossey-Bass, 989 Market Street, San Francisco, CA 94103-1741, USA

Wiley-VCH Verlag GmbH, Boschstr. 12, D-69469 Weinheim, Germany

John Wiley & Sons Australia Ltd, 33 Park Road, Milton, Queensland 4064, Australia

John Wiley & Sons (Asia) Pte Ltd, 2 Clementi Loop #02-01, Jin Xing Distripark, Singapore 129809

John Wiley & Sons Canada Ltd, 22 Worcester Road, Etobicoke, Ontario, Canada M9W 1L1

Wiley also publishes its books in a variety of electronic formats. Some content that appears in print may not be available in electronic books.

British Library Cataloguing in Publication Data

A catalogue record for this book is available from the British Library

ISBN 0-470-85163-5

This book is printed on acid-free paper responsibly manufactured from sustainable forestry in which at least two trees are planted for each one used for paper production.

Contents

Acknowledgments xi

Abbreviations xiii

Symbols xv

1 Introduction 1
   1.1   A Brief History 1
         1.1.1   *Early days* *1*
         1.1.2   *The relative decline of electric vehicles after 1910* *3*
         1.1.3   *Uses for which battery electric vehicles have remained popular*   *5*
   1.2   Developments Towards the End of the 20th Century 5
   1.3   Types of Electric Vehicle in Use Today 7
         1.3.1   *Battery electric vehicles* *8*
         1.3.2   *The IC engine/electric hybrid vehicle* *9*
         1.3.3   *Fuelled electric vehicles* *15*
         1.3.4   *Electric vehicles using supply lines* *18*
         1.3.5   *Solar powered vehicles* *18*
         1.3.6   *Electric vehicles which use flywheels or super capacitors* ... *18*
   1.4   Electric Vehicles for the Future 20
       Bibliography 21

2 Batteries 23
   2.1   Introduction 23
   2.2   Battery Parameters 24
         2.2.1   *Cell and battery voltages* *24*
         2.2.2   *Charge (or Amphour) capacity* *25*
         2.2.3   *Energy stored* *26*
         2.2.4   *Specific energy* *27*
         2.2.5   *Energy density* *27*
         2.2.6   *Specific power* *28*
         2.2.7   *Amphour (or charge) efficiency* *28*
         2.2.8   *Energy efficiency* *29*

		2.2.9	Self-discharge rates	29
		2.2.10	Battery geometry	29
		2.2.11	Battery temperature, heating and cooling needs	29
		2.2.12	Battery life and number of deep cycles	29
	2.3	Lead Acid Batteries		30
		2.3.1	Lead acid battery basics	30
		2.3.2	Special characteristics of lead acid batteries	32
		2.3.3	Battery life and maintenance	34
		2.3.4	Battery charging	35
		2.3.5	Summary of lead acid batteries	35
	2.4	Nickel-based Batteries		35
		2.4.1	Introduction	35
		2.4.2	Nickel cadmium	36
		2.4.3	Nickel metal hydride batteries	38
	2.5	Sodium-based Batteries		41
		2.5.1	Introduction	41
		2.5.2	Sodium sulphur batteries	41
		2.5.3	Sodium metal chloride (Zebra) batteries	42
	2.6	Lithium Batteries		44
		2.6.1	Introduction	44
		2.6.2	The lithium polymer battery	45
		2.6.3	The lithium ion battery	45
	2.7	Metal Air Batteries		46
		2.7.1	Introduction	46
		2.7.2	The aluminium air battery	46
		2.7.3	The zinc air battery	47
	2.8	Battery Charging		48
		2.8.1	Battery chargers	48
		2.8.2	Charge equalisation	49
	2.9	The Designer's Choice of Battery		51
		2.9.1	Introduction	51
		2.9.2	Batteries which are currently available commercially	52
	2.10	Use of Batteries in Hybrid Vehicles		53
		2.10.1	Introduction	53
		2.10.2	Internal combustion/battery electric hybrids	53
		2.10.3	Battery/battery electric hybrids	53
		2.10.4	Combinations using flywheels	54
		2.10.5	Complex hybrids	54
	2.11	Battery Modelling		54
		2.11.1	The purpose of battery modelling	54
		2.11.2	Battery equivalent circuit	55
		2.11.3	Modelling battery capacity	57
		2.11.4	Simulation a battery at a set power	61
		2.11.5	Calculating the Peukert Coefficient	64
		2.11.6	Approximate battery sizing	65

	2.12	In Conclusion	66
		References	67
3	**Alternative and Novel Energy Sources and Stores**		**69**
	3.1	Introduction	69
	3.2	Solar Photovoltaics	69
	3.3	Wind Power	71
	3.4	Flywheels	72
	3.5	Super Capacitors	74
	3.6	Supply Rails	77
		References	80
4	**Fuel Cells**		**81**
	4.1	Fuel Cells, a Real Option?	81
	4.2	Hydrogen Fuel Cells: Basic Principles	83
		4.2.1 Electrode reactions	*83*
		4.2.2 Different electrolytes	*84*
		4.2.3 Fuel cell electrodes	*87*
	4.3	Fuel Cell Thermodynamics – an Introduction	89
		4.3.1 Fuel cell efficiency and efficiency limits	*89*
		4.3.2 Efficiency and the fuel cell voltage	*92*
		4.3.3 Practical fuel cell voltages	*94*
		4.3.4 The effect of pressure and gas concentration	*95*
	4.4	Connecting Cells in Series – the Bipolar Plate	96
	4.5	Water Management in the PEM Fuel Cell	101
		4.5.1 Introduction to the water problem	*101*
		4.5.2 The electrolyte of a PEM fuel cell	*101*
		4.5.3 Keeping the PEM hydrated	*104*
	4.6	Thermal Management of the PEM Fuel Cell	105
	4.7	A Complete Fuel Cell System	107
		References	109
5	**Hydrogen Supply**		**111**
	5.1	Introduction	111
	5.2	Fuel Reforming	113
		5.2.1 Fuel cell requirements	*113*
		5.2.2 Steam reforming	*114*
		5.2.3 Partial oxidation and autothermal reforming	*116*
		5.2.4 Further fuel processing: carbon monoxide removal	*117*
		5.2.5 Practical fuel processing for mobile applications	*118*
	5.3	Hydrogen Storage I: Storage as Hydrogen	119
		5.3.1 Introduction to the problem	*119*
		5.3.2 Safety	*120*
		5.3.3 The storage of hydrogen as a compressed gas	*120*
		5.3.4 Storage of hydrogen as a liquid	*122*

		5.3.5	Reversible metal hydride hydrogen stores	124
		5.3.6	Carbon nanofibres	126
		5.3.7	Storage methods compared	127
	5.4	Hydrogen Storage II: Chemical Methods		127
		5.4.1	Introduction	127
		5.4.2	Methanol	128
		5.4.3	Alkali metal hydrides	130
		5.4.4	Sodium borohydride	132
		5.4.5	Ammonia	135
		5.4.6	Storage methods compared	138
		References		138
6	**Electric Machines and their Controllers**			**141**
	6.1	The 'Brushed' DC Electric Motor		141
		6.1.1	Operation of the basic DC motor	141
		6.1.2	Torque speed characteristics	143
		6.1.3	Controlling the brushed DC motor	147
		6.1.4	Providing the magnetic field for DC motors	147
		6.1.5	DC motor efficiency	149
		6.1.6	Motor losses and motor size	151
		6.1.7	Electric motors as brakes	153
	6.2	DC Regulation and Voltage Conversion		155
		6.2.1	Switching devices	155
		6.2.2	Step-down or 'buck' regulators	157
		6.2.3	Step-up or 'boost' switching regulator	159
		6.2.4	Single-phase inverters	162
		6.2.5	Three-phase	165
	6.3	Brushless Electric Motors		166
		6.3.1	Introduction	166
		6.3.2	The brushless DC motor	167
		6.3.3	Switched reluctance motors	169
		6.3.4	The induction motor	173
	6.4	Motor Cooling, Efficiency, Size and Mass		175
		6.4.1	Improving motor efficiency	175
		6.4.2	Motor mass	177
	6.5	Electrical Machines for Hybrid Vehicles		179
		References		181
7	**Electric Vehicle Modelling**			**183**
	7.1	Introduction		183
	7.2	Tractive Effort		184
		7.2.1	Introduction	184
		7.2.2	Rolling resistance force	184
		7.2.3	Aerodynamic drag	185
		7.2.4	Hill climbing force	185

	7.2.5	Acceleration force	185
	7.2.6	Total tractive effort	187
7.3	Modelling Vehicle Acceleration		188
	7.3.1	Acceleration performance parameters	188
	7.3.2	Modelling the acceleration of an electric scooter	189
	7.3.3	Modelling the acceleration of a small car	193
7.4	Modelling Electric Vehicle Range		196
	7.4.1	Driving cycles	196
	7.4.2	Range modelling of battery electric vehicles	201
	7.4.3	Constant velocity range modelling	206
	7.4.4	Other uses of simulations	207
	7.4.5	Range modelling of fuel cell vehicles	208
	7.4.6	Range modelling of hybrid electric vehicles	211
7.5	Simulations: a Summary		212
	References		212

8 Design Considerations — 213

8.1	Introduction	213
8.2	Aerodynamic Considerations	213
	8.2.1 Aerodynamics and energy	213
	8.2.2 Body/chassis aerodynamic shape	217
8.3	Consideration of Rolling Resistance	218
8.4	Transmission Efficiency	220
8.5	Consideration of Vehicle Mass	223
8.6	Electric Vehicle Chassis and Body Design	226
	8.6.1 Body/chassis requirements	226
	8.6.2 Body/chassis layout	227
	8.6.3 Body/chassis strength, rigidity and crash resistance	228
	8.6.4 Designing for stability	231
	8.6.5 Suspension for electric vehicles	231
	8.6.6 Examples of chassis used in modern battery and hybrid electric vehicles	232
	8.6.7 Chassis used in modern fuel cell electric vehicles	232
8.7	General Issues in Design	234
	8.7.1 Design specifications	234
	8.7.2 Software in the use of electric vehicle design	234

9 Design of Ancillary Systems — 237

9.1	Introduction	237
9.2	Heating and Cooling Systems	237
9.3	Design of the Controls	240
9.4	Power Steering	243
9.5	Choice of Tyres	243
9.6	Wing Mirrors, Aerials and Luggage Racks	243
9.7	Electric Vehicle Recharging and Refuelling Systems	244

10 Electric Vehicles and the Environment ... 245
- 10.1 Introduction ... 245
- 10.2 Vehicle Pollution: the Effects ... 245
- 10.3 Vehicles Pollution: a Quantitative Analysis ... 248
- 10.4 Vehicle Pollution in Context ... 251
- 10.5 Alternative and Sustainable Energy Used via the Grid ... 254
  - 10.5.1 Solar energy ... 254
  - 10.5.2 Wind energy ... 255
  - 10.5.3 Hydro energy ... 255
  - 10.5.4 Tidal energy ... 255
  - 10.5.5 Biomass energy ... 256
  - 10.5.6 Geothermal energy ... 257
  - 10.5.7 Nuclear energy ... 257
  - 10.5.8 Marine current energy ... 257
  - 10.5.9 Wave energy ... 257
- 10.6 Using Sustainable Energy with Fuelled Vehicles ... 258
  - 10.6.1 Fuel cells and renewable energy ... 258
  - 10.6.2 Use of sustainable energy with conventional IC engine vehicles ... 258
- 10.7 The Role of Regulations and Law Makers ... 258
- References ... 260

11 Case Studies ... 261
- 11.1 Introduction ... 261
- 11.2 Rechargeable Battery Vehicles ... 261
  - 11.2.1 Electric bicycles ... 261
  - 11.2.2 Electric mobility aids ... 263
  - 11.2.3 Low speed vehicles ... 263
  - 11.2.4 Battery powered cars and vans ... 266
- 11.3 Hybrid Vehicles ... 269
  - 11.3.1 The Honda Insight ... 269
  - 11.3.2 The Toyota Prius ... 271
- 11.4 Fuel Cell Powered Bus ... 272
- 11.5 Conclusion ... 275
- References ... 277

Appendices: MATLAB® Examples ... 279
- Appendix 1: Performance Simulation of the GM EV1 ... 279
- Appendix 2: Importing and Creating Driving Cycles ... 280
- Appendix 3: Simulating One Cycle ... 282
- Appendix 4: Range Simulation of the GM EV1 Electric Car ... 284
- Appendix 5: Electric Scooter Range Modelling ... 286
- Appendix 6: Fuel Cell Range Simulation ... 288
- Appendix 7: Motor Efficiency Plots ... 290

Index ... 293

Acknowledgments

The topic of electric vehicles is rather more interdisciplinary than a consideration of ordinary internal combustion engine vehicles. It covers many aspects of science and engineering. This is reflected in the diversity of companies that have helped with advice, information and pictures for this book. The authors would like to put on record their thanks to the following companies and organisations that have made this book possible.

    Ballard Power Systems Inc., Canada
    DaimlerChrysler Corp., USA and Germany
    The Ford Motor Co., USA
    General Motors Corp., USA
    GfE Metalle und Materialien GmbH, Germany
    Groupe Enerstat Inc., Canada
    Hawker Power Systems Inc., USA
    The Honda Motor Co. Ltd.
    Johnson Matthey Plc., UK
    MAN Nutzfahrzeuge AG, Germany
    MES-DEA SA, Switzerland
    Micro Compact Car Smart GmbH
    National Motor Museum Beaulieu
    Parry People Movers Ltd., UK
    Paul Scherrer Institute, Switzerland
    Peugeot S.A., France
    Powabyke Ltd., UK
    Richens Mobility Centre, Oxford, UK
    Saft Batteries, France
    SR Drives Ltd., UK
    Toyota Motor Co. Ltd.
    Wamfler GmbH, Germany
    Zytek Group Ltd., UK

In addition we would like to thank friends and colleagues who have provided valuable comments and advice. We are also indebted to these friends and colleagues, and our families, who have helped and put up with us while we devoted time and energy to this project.

                          James Larminie, Oxford Brookes University, Oxford, UK
                                   John Lowry, Acenti Designs Ltd., UK

Abbreviations

AC	Alternating current
BLDC	Brushless DC (motor)
BOP	Balance of plant
CARB	California air resources board
CCGT	Combined cycle gas turbine
CNG	Compressed natural gas
CPO	Catalytic partial oxidation
CVT	Continuously variable transmission
DC	Direct current
DMFC	Direct methanol fuel cell
ECCVT	Electronically controlled continuous variable transmission
ECM	Electronically commutated motor
EMF	Electromotive force
EPA	Environmental protection agency
EPS	Electric power steering
ETSU	Energy technology support unit (a government organisation in the UK)
EUDC	Extra-urban driving cycles
EV	Electric vehicle
FCV	Fuel cell vehicle
FHDS	Federal highway driving schedule
FUDS	Federal urban driving schedule
GM	General Motors
GM EV1	General Motors electric vehicle 1
GNF	Graphitic nanofibre
GTO	Gate turn off
HEV	Hybrid electric vehicle
HHV	Higher heating value
IC	Internal combustion
ICE	Internal combustion engine
IEC	International Electrotechnical Commission
IGBT	Insulated gate bipolar transistor
IMA	Integrated motor assist
IPT	Inductive power transfer

kph	Kilometres per hour
LHV	Lower heating value
LH_2	Liquid (cryogenic) hydrogen
LPG	Liquid petroleum gas
LSV	Low speed vehicle
MeOH	Methanol
mph	Miles per hour
MEA	Membrane electrode assembly
MOSFET	Metal oxide semiconductor field effect transistor
NASA	National Aeronautics and Space Administration
NiCad	Nickel cadmium (battery)
NiMH	Nickel metal hydride (battery)
NL	Normal litre, 1 litre at NTP
NTP	Normal temperature and pressure (20°C and 1 atm or 1.01325 bar)
NOX	Nitrous oxides
OCV	Open circuit voltage
PEM	Proton exchange membrane or polymer electrolyte membrane: different names for the same thing which fortunately have the same abbreviation
PEMFC	Proton exchange membrane fuel cell or polymer electrolyte membrane fuel cell
PM	Permanent magnet or particulate matter
POX	Partial oxidation
ppb	Parts per billion
ppm	Parts per million
PROX	Preferential oxidation
PWM	Pulse width modulation
PZEV	Partial zero emission vehicle
SAE	Society of Automotive Engineers
SFUDS	Simplified federal urban driving schedule
SL	Standard litre, 1 litre at STP
SOFC	Solid oxide fuel cell
SRM	Switched reluctance motor
STP	Standard temperature and pressure (= SRS)
SULEV	Super ultra low emission vehicles
TEM	Transmission electron microscope
ULEV	Ultra low emission vehicle
VOC	Volatile organic compounds
VRLA	Valve regulated (sealed) lead acid (battery)
WTT	Well to tank
WTW	Well to wheel
WOT	Wide open throttle
ZEBRA	Zero emissions battery research association
ZEV	Zero emission vehicle

Symbols

Letters are used to stand for variables, such as mass, and also as chemical symbols in chemical equations. The distinction is usually clear from the context, but for even greater clarity italics are use for variables, and ordinary text for chemical symbols, so H stands for enthalpy, whereas H stands for hydrogen.

In cases where a letter can stand for two or more variables, the context always makes it clear which is intended.

a	Acceleration
A	Area
B	Magnetic field strength
C_d	Drag coefficient
C	Amphour capacity of a battery OR capacitance of a capacitor
C_3	Amphour capacity of a battery if discharged in 3 hours, the '3 hour rate'
C_p	Peukert capacity of a battery, the same as the Amphour capacity if discharged at a current of 1 Amp
CR	Charge removed from a battery, usually in Amphours
CS	Charge supplied to a battery, usually in Amphours
d	Separation of the plates of a capacitor OR distance traveled
DoD	Depth of discharge, a ratio changing from 0 (fully charged) to 1 (empty)
E	Energy, or Young's modulus, or EMF (voltage)
E_b	Back EMF (voltage) of an electric motor in motion
E_s	Supplied EMF (voltage) to an electric motor
e	Magnitude of the charge on one electron, 1.602×10^{-19} Coulombs
f	Frequency
F	Force or Faraday constant, the charge on one mole of electrons, 96 485 Coulombs
F_{rr}	Force needed to overcome the rolling resistance of a vehicle
F_{ad}	Force needed to overcome the wind resistance on a vehicle
F_{la}	Force needed to give linear acceleration to a vehicle
F_{hc}	Force needed to overcome the gravitational force of a vehicle down a hill
$F_{\omega a}$	Force at the wheel needed to give rotational acceleration to the rotating parts of a vehicle
F_{te}	Tractive effort, the forward driving force on the wheels
g	Acceleration due to gravity

G	Gear ratio OR rigidity modulus OR Gibbs free energy (negative thermodynamic potential)
H	Enthalpy
I	Current, OR moment of inertia, OR second moment of area, the context makes it clear
I_m	Motor current
J	Polar second moment of area
k_c	Copper losses coefficient for an electric motor
k_i	Iron losses coefficient for an electric motor
k_w	Windage losses coefficient for an electric motor
KE	Kinetic energy
K_m	Motor constant
k	Peukert coefficient
L	Length
m	Mass
$\dot{m}$	Mass flow rate
m_b	Mass of batteries
N	Avogadro's number, 6.022×10^{23} OR revolutions per second
n	Number of cells in a battery, OR a fuel cell stack, OR the number of moles of substance
P	Power OR pressure
P_{adw}	Power at the wheels needed to overcome the wind resistance on a vehicle
P_{adb}	Power from the battery needed to overcome the wind resistance on a vehicle
P_{hc}	Power needed to overcome the gravitational force of a vehicle down a hill
P_{mot-in}	Electrical power supplied to an electric motor
$P_{mot-out}$	Mechanical power given out by an electrical motor
P_{rr}	Power needed to overcome the rolling resistance of a vehicle
P_{te}	Power supplied at the wheels of a vehicle
Q	Charge, e.g. in a capacitor
q	Sheer stress
R	Electrical resistance, OR the molar gas constant $8.314 \, \text{JK}^{-1} \, \text{mol}^{-1}$
R_a	Armature resistance of a motor or generator
R_L	Resistance of a load
r	Radius, of wheel, axle, OR the rotor of a motor, etc.
r_i, r_o	Inner and outer radius of a hollow tube
S	Entropy
SE	Specific energy
T	Temperature, OR Torque, OR the discharge time of a battery in hours
T_1, T_2	Temperatures at different stages in a process
T_f	Frictional torque, e.g. in an electrical motor
t_{on}, t_{off}	On and off times for a chopper circuit
v	Velocity
V	Voltage

W	Work done
z	Number of electrons transferred in a reaction
Φ	Total magnetic flux
δ	Deflection
δt	Time step in an iterative process
Δ	Change in ..., e.g. ΔH = change in enthalpy
σ	Bending stress
ε	Electrical permittivity
η	Efficiency
η_c	Efficiency of a DC/DC converter
η_{fc}	Efficiency of a fuel cell
η_m	Efficiency of an electric motor
η_g	Efficiency of a gearbox
η_0	Overall efficiency of a drive system
θ	Angle of deflection or bend
λ	Stoichiometric ratio
μ_{rr}	Coefficient of rolling resistance
ρ	Density
ψ	Angle of slope or hill
ω	Angular velocity

1

Introduction

The first demonstration electric vehicles were made in the 1830s, and commercial electric vehicles were available by the end of the 19th century. The electric vehicle has now entered its third century as a commercially available product and as such it has been very successful, outlasting many other technical ideas that have come and gone. However, electric vehicles have not enjoyed the enormous success of internal combustion (IC) engine vehicles that normally have much longer ranges and are very easy to refuel. Today's concerns about the environment, particularly noise and exhaust emissions, coupled to new developments in batteries and fuel cells may swing the balance back in favour of electric vehicles. It is therefore important that the principles behind the design of electric vehicles, the relevant technological and environmental issues are thoroughly understood.

1.1 A Brief History

1.1.1 Early days

The first electric vehicles of the 1830s used non-rechargeable batteries. Half a century was to elapse before batteries had developed sufficiently to be used in commercial electric vehicles. By the end of the 19th century, with mass production of rechargeable batteries, electric vehicles became fairly widely used. Private cars, though rare, were quite likely to be electric, as were other vehicles such as taxis. An electric New York taxi from about 1901 is shown, with Lily Langtree alongside, in Figure 1.1. Indeed if performance was required, the electric cars were preferred to their internal combustion or steam powered rivals. Figure 1.2 shows the first car to exceed the 'mile a minute' speed (60 mph) when the Belgium racing diver Camille Jenatzy, driving the electric vehicle known as 'La Jamais Contente',[1] set a new land speed record of 106 kph (65.7 mph). This also made it the first car to exceed 100 kph.

At the start of the 20th century electric vehicles must have looked a strong contender for future road transport. The electric vehicle was relatively reliable and started instantly,

---

[1] 'Ever striving' would be a better translation of this name, rather than the literal 'never happy'.

Electric Vehicle Technology Explained James Larminie and John Lowry
© 2003 John Wiley & Sons, Ltd ISBN: 0-470-85163-5

Figure 1.1 New York Taxi Cab in about 1901, a battery electric vehicle (The lady in the picture is Lillie Langtry, actress and mistress of King Edward VII.) (Photograph reproduced by permission of National Motor Museum Beaulieu.)

Figure 1.2 Camille Jenatzy's 'La Jamais Contente'. This electric car held the world land speed record in 1899, and was the first vehicle to exceed both 60 mph and 100 kph

whereas internal combustion engine vehicles were at the time unreliable, smelly and needed to be manually cranked to start. The other main contender, the steam engine vehicle, needed lighting and the thermal efficiency of the engines was relatively low.

By the 1920s several hundred thousand electric vehicles had been produced for use as cars, vans, taxis, delivery vehicles and buses. However, despite the promise of the early electric vehicles, once cheap oil was widely available and the self starter for the internal combustion engine (invented in 1911) had arrived, the IC engine proved a more attractive option for powering vehicles. Ironically, the main market for rechargeable batteries has since been for starting IC engines.

1.1.2 The relative decline of electric vehicles after 1910

The reasons for the greater success to date of IC engine vehicles are easily understood when one compares the specific energy of petroleum fuel to that of batteries. The specific energy[2] of fuels for IC engines varies, but is around $9000\,\text{Whkg}^{-1}$, whereas the specific energy of a lead acid battery is around $30\,\text{Whkg}^{-1}$. Once the efficiency of the IC engine, gearbox and transmission (typically around 20%) for a petrol engine is accounted for, this means that $1800\,\text{Whkg}^{-1}$ of useful energy (at the gearbox shaft) can be obtained from petrol. With an electric motor efficiency of 90% only $27\,\text{Whkg}^{-1}$ of useful energy (at the motor shaft) can be obtained from a lead acid battery. To illustrate the point further, 4.5 litres (1 gallon[3]) of petrol with a mass of around 4 kg will give a typical motor car a range of 50 km. To store the same amount of useful electric energy requires a lead acid battery with a mass of about 270 kg. To double the energy storage and hence the range of the petrol engine vehicle requires storage for a further 4.5 litres of fuel with a mass of around 4 kg only, whereas to do the same with a lead acid battery vehicle requires an additional battery mass of about 270 kg.

This is illustrated in Figure 1.3. In practice this will not double the electric vehicle range, as a considerable amount of the extra energy is needed to accelerate and decelerate the 270 kg of battery and to carry it up hills. Some of this energy may be regained through regenerative breaking, a system where the motor acts as a generator, braking the vehicle and converting the kinetic energy of the vehicle to electrical energy, which is returned to battery storage, from where it can be reused. In practice, when the efficiency of generation, control, battery storage and passing the electricity back through the motor and controller is accounted for, less than a third of the energy is likely to be recovered. As a result regenerative breaking tends to be used as much as a convenient way of braking heavy vehicles, which electric cars normally are, as for energy efficiency. For lead acid batteries to have the effective energy capacity of 45 litres (10 gallons) of petrol, a staggering 2.7 tonnes of batteries would be needed!

Another major problem that arises with batteries is the time it takes to recharge them. Even when adequate electrical power is available there is a minimum time, normally several hours, required to re-charge a lead acid battery, whereas 45 litres of petrol can be put into a vehicle in approximately one minute. The recharge time of some of the new batteries has been reduced to one hour, but this is still considerably longer than it takes to fill a tank of petrol.

Yet another limiting parameter with electric vehicles is that batteries are expensive, so that any battery electric vehicle is likely not only to have a limited range but to be more expensive than an internal combustion engine vehicle of similar size and build quality.

---

[2] 'Specific energy' means the energy stored per kilogram. The normal SI unit is Joule per kilogram (Jkg^{-1}). However, this unit is too small in this context, and so the Watthour per kilogram (Whkg^{-1}) is used instead. $1\,\text{Wh} = 3600\,\text{J}$.
[3] British gallon. In the USA a gallon is 3.8 litres.

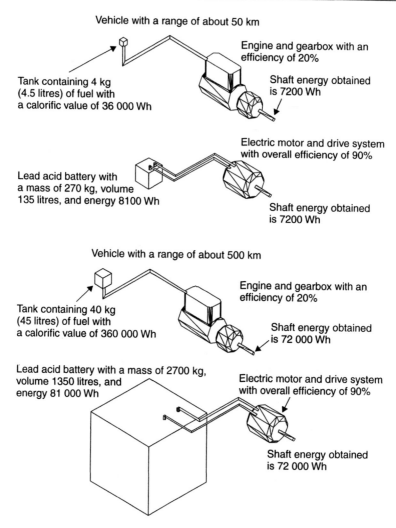

Figure 1.3 Comparison of energy from petrol and lead acid battery

For example, the 2.7 tonnes of lead acid batteries which give the same effective energy storage as 45 litres (10 UK gallons) of petrol would cost around £8000 at today's prices. The batteries also have a limited life, typically 5 years, which means that a further large investment is needed periodically to renew the batteries

When one takes these factors into consideration the reasons for the predominance of IC engine vehicles for most of the 20th Century become clear.

Since the 19th century ways of overcoming the limited energy storage of batteries have been used. The first is supplying the electrical energy via supply rails, the best example being the trolley bus. This has been widely used during the 20th century and allows quiet non-polluting buses to be used in towns and cities. When away from the electrical supply lines the buses can run from their own batteries. The downside is, of course, the expensive rather ugly supply lines which are needed and most trams and trolley bus systems have been removed from service. Modern inductive power transfer systems may overcome this problem.

Early on in the development of electric vehicles the concept was developed of the hybrid vehicle, in which an internal combustion engine driving a generator is used in conjunction with one or more electric motors. These were tried in the early 20th century, but recently have very much come back to the fore. The hybrid car is one of the most promising ideas which could revolutionise the impact of electric vehicles. The Toyota Prius (as in Figure 1.11) is a modern electric hybrid that, it is said, has more than doubled the number of electric cars on the roads. There is considerable potential for the development of electric hybrids and the idea of a hybrid shows considerable promise for future development. These are further considered in Section 1.3.2 below.

1.1.3 Uses for which battery electric vehicles have remained popular

Despite the above problems there have always been uses for electric vehicles since the early part of the 20th century. They have certain advantages over combustion engines, mainly that they produce no exhaust emissions in their immediate environment, and secondly that they are inherently quiet. This makes the electric vehicle ideal for environments such as warehouses, inside buildings and on golf courses, where pollution and noise will not be tolerated.

One popular application of battery/electric drives is for mobility devices for the elderly and physically handicapped. Indeed, in Europe and the United States the type of vehicle shown in Figure 1.4 is one of the most common. It can be driven on pavements, into shops, and in many buildings. Normally a range of 4 miles is quite sufficient but longer ranges allow disabled people to drive along country lanes. Another vehicle of this class is shown in Figure 11.2 of the final chapter.

They also retain their efficiencies in start-stop driving, when an internal combustion engine becomes very inefficient and polluting. This makes electric vehicles attractive for use as delivery vehicles such as the famous British milk float. In some countries this has been helped by the fact that leaving an unattended vehicle with the engine running, for example when taking something to the door of a house, is illegal.

1.2 Developments Towards the End of the 20th Century

During the latter part of the 20th century there have been changes which may make the electric vehicle a more attractive proposition. Firstly there are increasing concerns about the environment, both in terms of overall emissions of carbon dioxide and also the local emission of exhaust fumes which help make crowded towns and cities unpleasant to live in. Secondly there have been technical developments in vehicle design and improvements to rechargeable batteries, motors and controllers. In addition batteries which can be refueled and fuel cells, first invented by William Grove in 1840, have been developed to the point where they are being used in electric vehicles.

Figure 1.4 Electric powered wheel chair

Environmental issues may well be the deciding factor in the adoption of electric vehicles for town and city use. Leaded petrol has already been banned, and there have been attempts in some cities to force the introduction of zero emission vehicles. The state of California has encouraged motor vehicle manufacturers to produce electric vehicles with its Low Emission Vehicle Program. The fairly complex nature of the regulations in this state has led to very interesting developments in fuel cell, battery, and hybrid electric vehicles. (The important results of the Californian legislative programme are considered further in Chapter 10.)

Electric vehicles do not necessarily reduce the overall amount of energy used, but they do away with onboard generated power from IC engines fitted to vehicles and transfer the problem to the power stations, which can use a wide variety of fuels and where the exhaust emissions can be handled responsibly. Where fossil fuels are burnt for supplying electricity the overall efficiency of supplying energy to the car is not necessarily much better than using a diesel engine or the more modern highly efficient petrol engines. However there is more flexibility in the choice of fuels at the power stations. Also some or all the energy can be obtained from alternative energy sources such as hydro, wind or tidal, which would ensure overall zero emission.

Of the technical developments, the battery is an area where there have been improvements, although these have not been as great as many people would have wished. Commercially available batteries such nickel cadmium or nickel metal hydride can carry at best about double the energy of lead acid batteries, and the high temperature Sodium nickel chloride or Zebra battery nearly three times. This is a useful improvement, but still does not allow the design of vehicles with a long range. In practice, the available rechargeable battery with the highest specific energy is the lithium polymer battery which has a specific energy about three times that of lead acid. This is still expensive although there are signs that the price will fall considerably in the future. Zinc air batteries have potentially seven times the specific energy of lead acid batteries and fuel cells show considerable promise. So, for example, to replace the 45 litres (10 gallons) of petrol which would give a vehicle a range of 450 km (300 miles), a mass 800 kg of lithium battery would be required, an improvement on the 2700 kg mass of lead acid batteries, but still a large and heavy battery. Battery technology is addressed in much more detail in Chapter 2, and fuel cells are described in Chapter 4.

There have been increasing attempts to run vehicles from photovoltaic cells. Vehicles have crossed Australia during the World Solar Challenge with speeds in excess of 85 kph (50 mph) using energy entirely obtained from solar radiation. Although solar cells are expensive and of limited power ($100\,\text{Wm}^{-2}$ is typically achieved in strong sunlight), they may make some impact in the future. The price of photovoltaic cells is constantly falling, whilst the efficiency is increasing. They may well become useful, particularly for recharging commuter vehicles and as such are worthy of consideration.

1.3 Types of Electric Vehicle in Use Today

Developments of ideas from the 19th and 20th centuries are now utilised to produce a new range of electric vehicles that are starting to make an impact.

There are effectively six basic types of electric vehicle, which may be classed as follows. Firstly there is the traditional battery electric vehicle, which is the type that usually springs to mind when people think of electric vehicles. However, the second type, the hybrid electric vehicle, which combines a battery and an IC engine, is very likely to become the most common type in the years ahead. Thirdly there are vehicles which use replaceable fuel as the source of energy using either fuel cells or metal air batteries. Fourthly there are vehicles supplied by power lines. Fifthly there are electric vehicles which use energy directly from solar radiation. Sixthly there are vehicles that store energy by alternative means such as flywheels or super capacitors, which are nearly always hybrids using some other source of power as well.

Other vehicles that could be mentioned are railway trains and ships, and even electric aircraft. However, this book is focused on autonomous wheeled vehicles.

1.3.1 Battery electric vehicles

The concept of the battery electric vehicle is essentially simple and is shown in Figure 1.5. The vehicle consists of an electric battery for energy storage, an electric motor, and a controller. The battery is normally recharged from mains electricity via a plug and a battery charging unit that can either be carried onboard or fitted at the charging point. The controller will normally control the power supplied to the motor, and hence the vehicle speed, in forward and reverse. This is normally known as a 2 quadrant controller, forwards and backwards. It is usually desirable to use regenerative braking both to recoup energy and as a convenient form of frictionless braking. When in addition the controller allows regenerative braking in forward and reverse directions it is known as a 4 quadrant controller.[4]

There is a range of electric vehicles of this type currently available on the market. At the simplest there are small electric bicycles and tricycles and small commuter vehicles. In the leisure market there are electric golf buggies. There is a range of full sized electric vehicles, which include electric cars, delivery trucks and buses. Among the most important are also aids to mobility, as in Figure 1.4 and Figure 11.2 (in the final chapter), and also delivery vehicles and electric bicycles. Some examples of typical electrical vehicles using rechargeable batteries are shown in Figures 1.6 to 1.9. All of these vehicles have a fairly

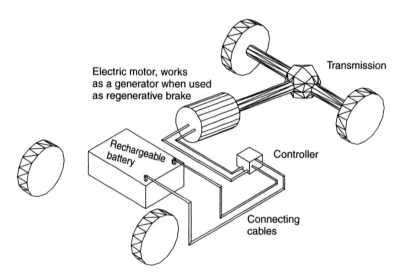

Figure 1.5 Concept of the rechargeable battery electric vehicle

---

[4] The 4 "quadrants" being forwards and backwards acceleration, and forwards and backwards braking.

Figure 1.6 The classic electric car, a battery powered city car (Picture of a Ford Th!nk® kindly supplied by the Ford Motor Co. Ltd.)

limited range and performance, but they are sufficient for the intended purpose. It is important to remember that the car is a very minor player in this field.

1.3.2 The IC engine/electric hybrid vehicle

A hybrid vehicle has two or more power sources, and there are a large number of possible variations. The most common types of hybrid vehicle combine an internal combustion engine with a battery and an electric motor and generator.

There are two basic arrangements for hybrid vehicles, the series hybrid and the parallel hybrid, which are illustrated in Figures 1.9 and 1.10 In the series hybrid the vehicle is driven by one or more electric motors supplied either from the battery, or from the IC engine driven generator unit, or from both. However, in either case the driving force comes entirely from the electric motor or motors.

In the parallel hybrid the vehicle can either be driven by the IC engine working directly through a transmission system to the wheels, or by one or more electric motors, or by both the electric motor and the IC engine at once.

In both series and parallel hybrids the battery can be recharged by the engine and generator while moving, and so the battery does not need to be anything like as large as in a pure battery vehicle. Also, both types allow for regenerative braking, for the drive motor to work as a generator and simultaneously slow down the vehicle and charge the battery.

Figure 1.7 Electric bicycles are among the most widely used electric vehicles

The series hybrid tends to be used only in specialist applications. For example, the diesel powered railway engine is nearly always a series hybrid, as are some ships. Some special all-terrain vehicles are series hybrid, with a separately controlled electric motor in each wheel. The main disadvantage of the series hybrid is that all the electrical energy must pass through both the generator and the motors. The adds considerably to the cost of such systems.

The parallel hybrid, on the other hand, has scope for very wide application. The electric machines can be much smaller and cheaper, as they do not have to convert all the energy.

Figure 1.8 Delivery vehicles have always been an important sector for battery powered electric vehicles

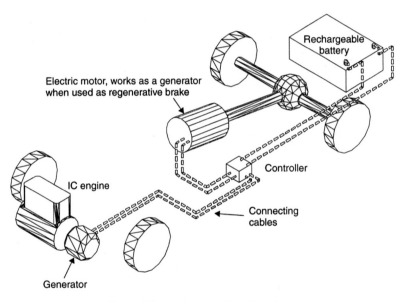

Figure 1.9 Series hybrid vehicle layout

There are various ways in which a parallel hybrid vehicle can be used. In the simplest it can run on electricity from the batteries, for example, in a city where exhaust emissions are undesirable, or it can be powered solely by the IC engine, for example, when traveling outside the city. Alternatively, and more usefully, a parallel hybrid vehicle can use the IC engine and batteries in combination, continually optimising the efficiency of the IC

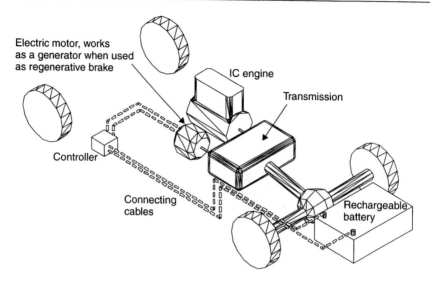

Figure 1.10 Parallel hybrid vehicle layout

engine. A popular arrangement is to obtain the basic power to run the vehicle, normally around 50% of peak power requirements, from the IC engine, and to take additional power from the electric motor and battery, recharging the battery from the engine generator when the battery is not needed. Using modern control techniques the engine speed and torque can be controlled to minimise exhaust emissions and maximise fuel economy. The basic principle is to keep the IC engine working fairly hard, at moderate speeds, or else turn it off completely.

In parallel hybrid systems it is useful to define a variable called the 'degree of hybridisation' as follows:

$$\text{DOH} = \frac{\text{electric motor power}}{\text{electric motor power} + \text{IC engine power}}$$

The greater the degree of hybridisation, the greater the scope for using a smaller IC engine, and have it operating at near its optimum efficiency for a greater proportion of the time. At the time of writing the highly important California Air Resources Board (CARB) identifies three levels of hybridisation, as in Table 1.1. The final row gives an indication of the perceived 'environmental value' of the car, and issue considered in Chapter 10.

Because there is the possibility of hybrid vehicles moving, albeit for a short time, with the IC engine off and entirely under battery power, they can be called 'partial zero emission vehicles' (PZEVs).

Hybrid vehicles are more expensive than conventional vehicles. However there are some savings which can be made. In the series arrangement there is no need for a gear box, transmission can be simplified and the differential can be eliminated by using a pair of motors fitted on opposite wheels. In both series and parallel arrangements the conventional battery starter arrangement can be eliminated.

Introduction

Table 1.1 CARB classification of hybrid electric vehicles, as in April 2003

	Level 1: low-voltage HEV	Level 2: high-voltage HEV	Level 3: high-voltage high-power HEV
Motor drive voltage	<60 V	≥60 V	≥60 V
Motor drive peak power	≥4 kW	≥10 kW	≥50 kW
Regenerative braking	Yes	Yes	Yes
Idle stop/start	Yes	Yes	Yes
10 year/150 kmile battery warranty	Yes	Yes	Yes
ZEV program credit	0.2	0.4	0.5

Figure 1.11 The Toyota Prius (Pictures reproduced by kind permission of Toyota.)

There are several hybrid vehicles currently on the market, and this is a sector that is set to grow rapidly in the years ahead. The Toyota Prius, shown in Figure 1.11, is the vehicle which really brought hybrid vehicles to public attention. Within two years of its launch in 1998 it more than doubled the number of electric vehicles on the roads of Japan.[5] The Prius uses a 1.5 litre petrol engine and a 33 kW electric motor either in combination or separately to produce the most fuel-efficient performance. A nickel metal hydride battery is used. At start up or at low speeds the Prius is powered solely by the electric motor, avoiding the use of the internal combustion engine when it is at its most polluting and least efficient. This car uses regenerative braking and has a high overall fuel economy of about 56.5 miles per US gallon (68 miles per UK gallon).[6] The Prius has a top speed of 160 km/h (100 mph) and accelerates to 100 km/h (62 mph) in 13.4 seconds. The Prius battery is only charged from the engine and does not use an external socket. It is therefore refueled with petrol only, in the conventional way. In addition, it seats four people in comfort, and the luggage space is almost unaffected by the somewhat larger than normal battery. The fully automatic transmission system is a further attraction of this car that

---

[5] Honda brought out its parallel hybrid Insight model in 1998. This has somewhat better fuel economy and lower emissions. However, it is only a two-seater, the luggage space is much more limited, and its market impact has not been so great.

[6] Further details in Table 11.6 of the final chapter.

has put electric cars well into the realm of the possible for ordinary people making the variety of journeys they expect their cars to cope with.

The Toyota Prius mainly has the characteristics of a parallel hybrid, as in Figure 1.10, in that the IC engine can directly power the vehicle. However, it does have a separate motor and generator, can operate in series mode, and is not a 'pure' parallel hybrid. It has a fairly complex 'power splitter' gearbox, based on epicyclic gears, that allows power from both the electric motor or the IC engine, in almost any proportion, to be sent to the wheels or gearbox. Power can also be sent from the wheels to the generator for regenerative braking.

Most of the major companies are now bringing out vehicles that are true parallel hybrids. The Honda Insight, shown in Figure 8.14, and whose performance figures are given in Table 11.5, is a good example. There is also now a parallel hybrid electric version of the popular Honda Civic available.

As well as the parallel hybrid arrangement shown in Figure 1.10, in which the IC engine and electric machine sit side by side, there is an almost infinite number of other possible arrangements. The Honda vehicles mentioned above have the electric machine sitting in line with the crankshaft, in the place of the flywheel in a conventional IC engine. Other notable hybrids appearing on the market, such as hybrid versions of the popular sports utility vehicle (SUV) in the USA, have the IC engine and the electric machines connected to different axles, as in Figure 1.12. Here the electric system drives the rear wheels, and the IC engine the front. This is a true parallel hybrid, and the road can be thought of as the medium that connects the two parts of the system, electric and engine. This arrangement has many attractions in terms of simplicity of packaging, and that it is a very neat way of giving the vehicle a four wheel drive capability. The battery will be mainly charged by regenerative braking, but if that is insufficient, at times of low speed travel the rear wheels could be electrically braked thus charging the battery, and the front driven harder to maintain speed. This transfers energy from the IC engine to the battery, using the road.

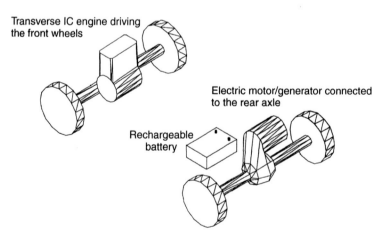

Figure 1.12 Parallel hybrid system with IC engine driving the front axle, and electric power to the rear wheels

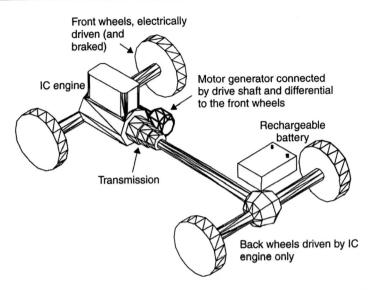

Figure 1.13 Yet another possible parallel hybrid arrangement: electric power to the front wheels, IC engine to the rear

A variation on this idea is shown in Figure 1.13. Many SUVs are rear wheel driven by the IC engine, and these can be made into parallel hybrids by providing electrical power to the *front* axle. This has a slight advantage over the arrangement in Figure 1.12, in that more regenerative braking power is available on the front axle, due to the weight transfer to the front wheels under braking.

Even more variety in the arrangement for hybrid vehicles becomes apparent when we note that, with all types of hybrid, the battery could be charged from a separate electrical supply, such as the mains, while the vehicle is not in use. This would only be worthwhile if a larger battery was used, and this could allow the car considerable 'battery only' range. There are no manufacturers with vehicles of this type planned for launch soon, but it might be a development in years ahead.

Despite the huge variety in detail that is possible with IC engine/battery hybrids, the major technological components are essentially the same: electric motors, batteries, and controllers. So we do not have a chapter dedicated to hybrids, which their importance could justify, because the underlying technology is explained the chapters covering these topics.

1.3.3 Fuelled electric vehicles

The basic principle of electric vehicles using fuel is much the same as with the battery electric vehicle, but with a fuel cell or metal air battery replacing the rechargeable electric battery. Most of the major motor companies have developed very advanced fuel cell powered cars. Daimler Chrysler for example have developed fuel cell cars based on the Mercedes A series, fitted with Ballard fuel cells, one of which is shown in Figure 1.14. This fuel cell runs on hydrogen which is stored in liquid form.

Figure 1.14 The Necar 4 fuel cell car from 1999. This was the first fuel cell car to have a performance and range similar to IC engine vehicles. The top speed is 144 kph, and the range 450 km. The hydrogen fuel is stored as a liquid. (Photograph reproduced by kind permission of Ballard Power Systems.)

Although invented in about 1840, fuel cells are an unfamiliar technology for most people, and they are considered in some detail in Chapter 4.

As we shall see in later in Chapter 5, a major issue with fuel cells is that, generally, they require hydrogen fuel. This can be stored on board, though this is not easy. An alternative is to make the hydrogen from a fuel such as methanol. This is the approach taken with the Necar 5, a further development of the vehicle in Figure 1.14, which can simply be refuelled with methanol in the same way as a normal vehicle is filled up with petrol. The car has a top speed of 150 kph, an overall fuel consumption of 5 l/100 km of methanol. It is shown in Figure 5.5.

Another fuel cell vehicle of note is the Honda FCX shown in Figure 1.15, which was the first fuel cell vehicle in the USA to be registered officially as a zero emission vehicle (ZEV) with the environmental protection agency (EPA).

Public service vehicles such as buses can more conveniently use novel fuels such as hydrogen, because they only fill up at one place. Buses are a very promising early application of fuel cells, and an example is shown in Figure 1.16.

Zinc air batteries produced by the Electric Fuel Transportation Company have been tested in vehicles both in the USA and in Europe. The company's stated mission is to bring about the deployment of commercial numbers of zinc-air electric buses, in this decade. During the summer of 2001 a zero emission zinc-air transport bus completed tests at sites in New York State, and later in the year was demonstrated in Nevada. In Germany, a

Figure 1.15 The Honda FCX was the first fuel cell car to be certified for use by the general public in the USA in 2002, and so theoretically become publicly available. This four seater city car has a top speed of 150 kph and a range of 270 km. The hydrogen fuel is stored in a high-pressure tank (Photograph reproduced by kind permission of Ballard Power Systems.)

Figure 1.16 Citaro fuel cell powered bus, one of a fleet entering service in 2003 (Photograph reproduced by kind permission of Ballard Power Systems.)

government-funded consortium of industrial firms is developing a zero emission delivery vehicle based on EFTC's zinc air batteries.

Metal air batteries (described in Chapter 2) are a variation on fuel cells. They are refuelled by replacing the metal electrodes which can be recycled. Zinc air batteries are a particularly promising battery in this class.

1.3.4 Electric vehicles using supply lines

Both the trolley bus and the tram are well known, and at one time were widely used as a means of city transport. They are a cost effective, zero emission form of city transport that is still used in some cities. Normally electricity is supplied by overhead supply lines and a small battery is used on the trolley bus to allow it a limited range without using the supply lines.

It is now difficult to see why most of these have been withdrawn from service. It must be remembered that at the time when it became fashionable to remove trams and trolley buses from service, cost was a more important criterion than environmental considerations and worries about greenhouse gases. Fossil fuel was cheap and overhead wires were considered unsightly, inflexible, expensive and a maintenance burden. Trams in particular were considered to impede the progress of the all-important private motor car. Today, when IC engine vehicles are clogging up and polluting towns and cities, the criteria have changed again. Electric vehicles powered by supply lines could make a useful impact on modern transport and the concept should not be overlooked by designers, although most of this book is devoted to autonomous vehicles.

1.3.5 Solar powered vehicles

Solar powered vehicles such as the Honda Dream, which won the 1996 world solar challenge, are expensive and only work effectively in areas of high sunshine. The Honda Dream Solar car achieved average speeds across Australia, from Darwin to Adelaide, of 85 kph (50 mph). Although it is unlikely that a car of this nature would be a practical proposition as a vehicle for every day use, efficiencies of solar photovoltaic cells are rising all the time whilst their cost is decreasing. The concept of using solar cells, which can be wrapped to the surface of the car to keep the batteries of a commuter vehicle topped up, is a perfectly feasible idea, and as the cost falls and the efficiency increases may one day prove a practical proposition.

1.3.6 Electric vehicles which use flywheels or super capacitors

There have been various alternative energy storage devices including the flywheel and super capacitors. As a general rule both of these devices have high specific powers, which means that they can take in and give out energy very quickly. However, the amount of energy they can store is currently rather small. In other words, although they have a good *power* density, they have a poor *energy* density. These devices are considered in more detail in Chapter 3.

A novel electric vehicle using a flywheel as an energy storage device was designed by John Parry, UK. The vehicle is essentially a tram in which the flywheel is speeded up by an electric motor. Power to achieve this is supplied when the tram rests whilst picking up passengers at one of its frequent stations. The tram is driven from the flywheel by an infinitely variable cone and ball gearbox. The tram is decelerated by using the gearbox

Figure 1.17 The Parry People Mover. This electric vehicle uses a flywheel to store energy (Photograph kindly supplied by Parry People Movers Ltd.)

to accelerate the flywheel and hence transfer the kinetic energy of the vehicle to the kinetic energy of the flywheel, an effective form of regenerative braking. The vehicle is illustrated in Figure 1.17. The inventor has proposed fitting both the flywheel and gearbox to a conventional battery powered car. The advantage of this is that batteries do not readily take up and give out energy quickly, whereas a flywheel can. Secondly the arrangement can be made to give a reasonably high efficiency of regeneration, which will help to reduce the battery mass.

Experimental vehicles using ultra capacitors (also considered in Chapter 3) to store power have also been tested; normally they are used as part of a hybrid vehicle. The main source of power can be an IC engine, as with the bus shown in Figure 1.18, or it could be a fuel cell. The MAN bus in Figure 1.18 uses a diesel engine. In either case the purpose of the capacitor is to allow the recovery of kinetic energy when the vehicle slows down, and to increase the available peak power during times of rapid acceleration, thus allowing a smaller engine or fuel cell to power a vehicle.

Energy stores such as capacitors and flywheels can be used in a wide range of hybrids. Energy providers which can be used in hybrid vehicles include rechargeable batteries, fuelled batteries or fuel cells, solar power, IC engines, supply lines, flywheels and capacitors. Any two or more of these can be used together to form a hybrid electric vehicle, giving over 21 combinations of hybrids with 2 energy sources. If 3 or more energy sources are combined there are a further 35 combinations. Certainly there is plenty of scope for imagination in the use of hybrid combinations.

Figure 1.18 Hybrid diesel/electric bus, with electrical energy stored in capacitors (Photograph reproduced by kind permission of Man Nutzfahrzeuge AG.)

1.4 Electric Vehicles for the Future

The future of electric vehicles, of course, remains to be written. However, the need for vehicles that minimise the damage to the environment is urgent. Much of the technology to produce such vehicles has been developed and the cost, currently high in many cases, is likely to drop with increasing demand, which will allow quantity production.

The following chapters describe the key technologies that are the basis of electric vehicles now and in the future: batteries (Chapter 2), other energy stores such as capacitors and flywheels (Chapter 3), fuel cells (Chapter 4), hydrogen supply (Chapter 5), and electric motors (Chapter 6). Once the basic concepts are understood, their incorporation into vehicles can be addressed. A very important aspect of this is vehicle performance modelling, and so Chapter 7 is devoted to this topic. The final chapters address the important topics of the design of safe and stable vehicles, and of the 'comfort facilities' that are essential in a modern car. Finally, the environmental impact of electric vehicles needs to be honestly addressed; to what extent do they really reduce the environmental damage done by our love of personal mobility?

There is a real prospect of cities and towns using zero emission vehicles, and also of vehicles that use electrical technology to reduce fuel consumption. It is up to engineers, scientists and designers to make this a reality.

Bibliography

The following two books have good summaries of the history of electric vehicles.

Wakefield E.H. (1994) *History of the Electric Automobile*, The Society of Automobile Engineers, Warrendale.
Westbrook M.H. (2001) *The Electric Car*, The Institution of Electrical Engineers, London.

2

Batteries

2.1 Introduction

We have seen in the previous chapter that there are many different types and sizes of electric vehicle. However, in nearly all cases the battery is a key component. In the classic electrical vehicle the battery is the only energy store, and the component with the highest cost, weight and volume. In hybrid vehicles the battery, which must continually accept and give out electrical energy, is also a key component of the highest importance. Some fuel cell vehicles have been made which have batteries that are no larger than those normally fitted to internal combustion engined cars, but it is probable that most early FC powered vehicles will have quite large batteries and work in hybrid fuel cell/battery mode. In short, a good understanding of battery technology and performance is vital to anyone involved with electric vehicles.

What is an electric battery? A battery consists of two or more electric cells joined together. The cells convert chemical energy to electrical energy. The cells consist of positive and negative electrodes joined by an electrolyte. It is the chemical reaction between the electrodes and the electrolyte which generates DC electricity. In the case of secondary or rechargeable batteries, the chemical reaction can be reversed by reversing the current and the battery returned to a charged state.

The 'lead acid' battery is the most well known rechargeable type, but there are others. The first electric vehicle using rechargeable batteries preceded the invention of the rechargeable lead acid by quarter of a century, and there are a very large number of materials and electrolytes that can be combined to form a battery. However, only a relatively small number of combinations have been developed as commercial rechargeable electric batteries suitable for use in vehicles. At present these include lead acid, nickel iron, nickel cadmium, nickel metal hydride, lithium polymer and lithium iron, sodium sulphur and sodium metal chloride. There are also more recent developments of batteries that can be mechanically refuelled, the main ones being aluminium-air and zinc-air. Despite all the different possibilities tried, and about 150 years of development, a suitable battery has still not yet been developed which allows widespread use of electric vehicles. However, there have recently been some important developments in battery technology that hold out great hope for the future. Also, provided their performance is understood and properly

---

Electric Vehicle Technology Explained   James Larminie and John Lowry
© 2003 John Wiley & Sons, Ltd   ISBN: 0-470-85163-5

modelled, it is perfectly possible to design very useful vehicles using current batteries as the only or principal energy store.

From the electric vehicle designer's point of view the battery can be treated as a 'black box' which has a range of performance criteria. These criteria will include specific energy, energy density, specific power, typical voltages, amp hour efficiency, energy efficiency, commercial availability, cost, operating temperatures, self-discharge rates, number of life cycles and recharge rates, terms which will be explained in the following section. The designer also needs to understand how energy availability varies with regard to ambient temperature, charge and discharge rates, battery geometry, optimum temperature, charging methods, cooling needs and likely future developments. However, at least a basic understanding of the battery chemistry is very important, otherwise the performance and maintenance requirements of the different types, and most of the disappointments connected with battery use, such as their limited life, self-discharge, reduced efficiency at higher currents, and so on, cannot be understood. This basic knowledge is also needed in regard to likely hazards in an accident and the overall impact of the use of battery chemicals on the environment. Recycling of used batteries is also becoming increasingly important.

The main parameters that specify the behaviour and performance of a battery are given in the following section. In the later sections the chemistry and performance of the most important battery types are outlined, and finally the very important topic of battery performance modelling is outlined.

2.2 Battery Parameters

2.2.1 Cell and battery voltages

All electric cells have nominal voltages which gives the approximate voltage when the cell is delivering electrical power. The cells can be connected in series to give the overall voltage required. Traction batteries for electric vehicles are usually specified as 6 V or 12 V, and these units are in turn connected in series to produce the voltage required. This voltage will, in practice, change. When a current is given out, the voltage will fall; when the battery is being charged, the voltage will rise.

This is best expressed in terms of 'internal resistance', and the equivalent circuit of a battery is shown in Figure 2.1. The battery is represented as having a fixed voltage E, but the voltage at the terminals is a different voltage V, because of the voltage across the internal resistance R. Assuming that a current I is flowing out of the battery, as in Figure 2.1, then by basic circuit theory we can say that:

$$V = E - IR \qquad (2.1)$$

Note that if the current I is zero, the terminal voltage is equal to E, and so E is often referred to as the open circuit voltage. If the battery is being charged, then clearly the voltage will *increase* by IR. In electric vehicle batteries the internal resistance should clearly be as low as possible.[1]

---

[1] A good quality 12 V, 25 Amphour lead acid battery will typically have an internal resistance of about 0.005 ohms.

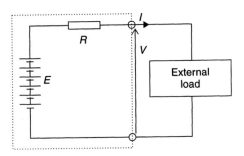

Figure 2.1 Simple equivalent circuit model of a battery. This battery is composed of six cells

Generally this equation (2.1) gives a fairly good prediction of the 'in use' battery voltage. However, the open circuit voltage E is not in fact constant. The voltage is also affected by the 'state of charge', and other factors such as temperature. This is dealt with in more detail in Section 2.11, where we address the problem of modelling the performance of batteries.

2.2.2 Charge (or Amphour) capacity

The electric charge that a battery can supply is clearly a most crucial parameter. The SI unit for this is the Coulomb, the charge when one Amp flows for one second. However, this unit is inconveniently small. Instead the Amphour is used: one Amp flowing for one hour. The capacity of a battery might be, say, 10 Amphours. This means it can provide 1 Amp for 10 hours, or 2 Amps for 5 hours, or in theory 10 Amps for 1 hour. However, in practice, it does not work out like this for most batteries.

It is usually the case that while a battery may be able to provide 1 Amp for 10 hours, if 10 Amps are drawn from it, it will last less than one hour. It is most important to understand this. The capacity of the large batteries used in electric vehicles (traction batteries) is usually quoted for a 5 hour discharge. Figure 2.2 shows how the capacity is affected if the charged is removed more quickly, or more slowly. The diagram is for a nominally 100 Amphour battery. Notice that if the charge is removed in one hour, the capacity falls very considerably to about 70 Amphours. On the other hand, if the current is drawn off more slowly, in say 20 hours, the capacity rises to about 110 Amphours.

This change in capacity occurs because of unwanted side reactions inside the cell. The effect is most noticeable in the lead acid battery, but occurs in all types. It is very important to be able to accurately predict the effects of this phenomenon, and that is addressed in Section 2.11, when we consider battery modelling.

The charge capacity leads to an important notation point that should be explained at this point. The capacity of a battery in Amphours is represented by the letter C. However, somewhat confusingly, until you get used to it, this is also used to represent a current.

Suppose a battery has a capacity of 42 Amphours, then it is said that $C = 42$ Amphours. Battery users talk about 'a discharge current of $2C$', or 'charging the battery at $0.4C$'. In these cases this would mean a discharge current of 84 Amps, or a charging current of 16.8 Amps.

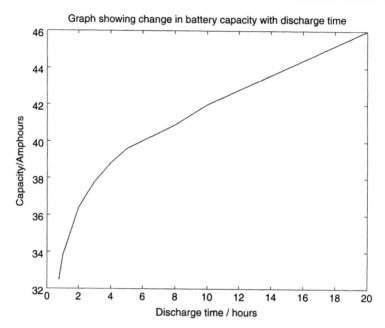

Figure 2.2 Graph showing the change in amphour charge capacity of a nominally 42 Amphour battery. This graph is based on measurements from a lead acid traction battery produced by Hawker Energy Products Inc

A further refinement is to give a subscript on the C symbol. As we noted above, the Amphour capacity of a battery varies with the time taken for the discharge. In our example, the 42 Amphour battery is rated thus for a 10 hour discharge. In this more complete notation, a discharge current of 84 Amps should be written as $2C_{10}$.

Example: Express the current 21 Amps from our example 42 Amphour battery, in C notation.

As a ratio of 42 Amps, 21 is 1/2 or 0.5. Thus the current 21 Amps $= 0.5C_{10}$.

This way of expressing a battery current is very useful, as it relates the current to the size of the battery. It is almost universally used in battery literature and specifications, though the subscript relating to the rated discharge time is often omitted.

2.2.3 Energy stored

The purpose of the battery is to store energy. The energy stored in a battery depends on its *voltage*, and the *charge* stored. The SI unit is the Joule, but this is an inconveniently small unit, and so we use the Watthour instead. This is the energy equivalent of working at a power of 1 Watt for 1 hour. The Watthour is equivalent to 3600 Joules. The Watthour is compatible with our use of the Amphour for charge, as it yields the simple formula:

$$\text{Energy in Watthours} = \text{Voltage} \times \text{Amphours or Energy} = V \times C \quad (2.2)$$

However, this equation must be used with great caution. We have noted that both the battery voltage V, and even more so the Amphour capacity C, vary considerably depending on how the battery is used. Both are reduced if the current is increased and the battery is drained quickly. The stored energy is thus a rather variable quantity, and reduces if the energy is released quickly. It is usually quoted in line with the Amphour rating, i.e. if the charge capacity is given for a five hour discharge, then the energy should logically be given for this discharge rate.

2.2.4 Specific energy

Specific energy is the amount of electrical energy stored for every kilogram of battery mass. It has units of $Wh.kg^{-1}$. Once the energy capacity of the battery needed in a vehicle is known (Wh) it can be divided by the specific energy ($Wh.kg^{-1}$) to give a first approximation of the battery mass. Specific energies quoted can be no more than a guide, because as we have seen, the energy stored in a battery varies considerably with factors such as temperature and discharge rate.

We will see in Section 2.2.6 below, and in the Ragone plot of Figure 2.3 how much the specific energy of a battery can change.

2.2.5 Energy density

Energy density is the amount of electrical energy stored per cubic metre of battery volume. It normally has units of $Wh.m^{-3}$. It is also an important parameter as the energy capacity

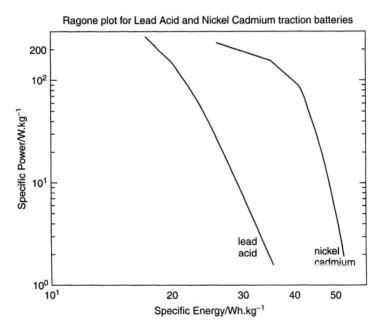

Figure 2.3 A Ragone plot – specific power versus specific energy graph – for typical lead acid and nickel cadmium traction batteries

of the battery (Wh) can be divided by the battery's energy density (Wh.m^{-3}) to show the volume of battery required. Alternatively if a known volume is available for batteries, the volume (m^3) can be multiplied by the batteries energy density (Wh.m^{-3}) to give a first approximation of how much electrical energy can be made available. The battery volume may well have a considerable impact on vehicle design. As with specific energy, the energy density is a nominal figure.

2.2.6 Specific power

Specific power is the amount of power obtained per kilogram of battery. It is a highly variable and rather anomalous quantity, since the power given out by the battery depends far more upon the load connected to it than the battery itself. Although batteries do have a maximum power, it is not sensible to operate them at anywhere near this maximum power for more than a few seconds, as they will not last long and would operate very inefficiently.

The normal units are W.kg^{-1}. Some batteries have a very good specific energy, but have low specific power, which means they store a lot of energy, but can only give it out slowly. In electric vehicle terms, they can drive the vehicle very slowly over a long distance. High specific power normally results in lower specific energy for any particular type of battery. This is because, as we saw in Section 2.2.2, taking the energy out of a battery quickly, i.e. at high power, reduces the energy available.

This difference in change of specific power with specific energy for different battery types is very important, and it is helpful to be able to compare them. This is often done using a graph of specific power against specific energy, which is known as a Ragone plot. Logarithmic scales are used, as the power drawn from a battery can vary greatly in different applications. A Ragone plot for a good quality lead acid traction battery, and a similar NiCad battery, is shown in Figure 2.3.

It can be seen that, for both batteries, as the specific power increases the specific energy is reduced. In the power range 1 to 100 W.kg^{-1} the NiCad battery shows slightly less change. However, above about 100 W.kg^{-1} the NiCad battery falls much faster than the lead acid.

Ragone plots like Figure 2.3 are used to compare energy sources of all types. In this case we should conclude that, ignoring other factors such as cost, the NiCad battery performs better if power densities of less than 100 W.kg^{-1} are required. However, at higher values, up to 250 W.kg^{-1} or more, then the lead acid begins to become more attractive. The Ragone plot also emphasises the point that a simple single number answer cannot be given to the question 'What is the specific power of this battery?'.

2.2.7 Amphour (or charge) efficiency

In an ideal world a battery would return the entire charge put into it, in which case the amp hour efficiency is 100%. However, no battery does; its charging efficiency is less than 100%. The precise value will vary with different types of battery, temperature and rate of charge. It will also vary with the state of charge. For example, when going from about 20% to 80% charged the efficiency will usually be very close to 100%, but as the last 20% of the charge is put in the efficiency falls off greatly. The reasons for this will be made clear when we look at each of the battery types later in the chapter.

2.2.8 Energy efficiency

This is another very important parameter and it is defined as the ratio of electrical energy supplied by a battery to the amount of electrical energy required to return it to the state before discharge. A strong argument for using electric vehicles is based on its efficient use of energy, with resulting reduction of overall emissions; hence high energy efficiency is desirable. It should be clear from what has been said in the preceding sections that the energy efficiency will vary greatly with how a battery is used. If the battery is charged and discharged rapidly, for example, energy efficiency decreases considerably. However it does act as a guide for comparing batteries, in much the same way as fuel consumption does for cars.

2.2.9 Self-discharge rates

Most batteries discharge when left unused, and this is known as self-discharge. This is important as it means some batteries cannot be left for long periods without recharging. The reasons for this self-discharge will be explained in the sections that follow. The rate varies with battery type, and with other factors such as temperature; higher temperatures greatly increase self-discharge.

2.2.10 Battery geometry

Cells come in many shapes: round, rectangular, prismatic or hexagonal. They are normally packaged into rectangular blocks. Some batteries can be supplied with a fixed geometry only. Some can be supplied in a wide variation of heights, widths and lengths. This can give the designer considerable scope, especially when starting with a blank sheet of paper, or more likely today a blank CAD screen. He/she could, for example, spread the batteries over the whole floor area, ensuring a low centre of gravity and very good handling characteristics.

2.2.11 Battery temperature, heating and cooling needs

Although most batteries run at ambient temperature, some run at higher temperatures and need heating to start with and then cooling when in use. In others, battery performance drops off at low temperatures, which is undesirable, but this problem could be overcome by heating the battery. When choosing a battery the designer needs to be aware of battery temperature, heating and cooling needs, and has to take these into consideration during the vehicle design process.

2.2.12 Battery life and number of deep cycles

Most rechargeable batteries will only undergo a few hundred deep cycles to 20% of the battery charge. However, the exact number depends on the battery type, and also on the details of the battery design, and on how the battery is used. This is a very important figure in a battery specification, as it reflects in the lifetime of the battery, which in turn reflects in electric vehicle running costs. More specific information about this, and all the other battery parameters mentioned, are given in the sections that follow on particular battery types.

2.3 Lead Acid Batteries

2.3.1 Lead acid battery basics

The best known and most widely used battery for electric vehicles is the lead acid battery. Lead acid batteries are widely used in IC engine vehicles and as such are well known. However for electric vehicles, more robust lead acid batteries that withstand deep cycling and use a gel rather than a liquid electrolyte are used. These batteries are more expensive to produce.

In the lead acid cells the negative plates have a spongy lead as their active material, whilst the positive plates have an active material of lead dioxide. The plates are immersed in an electrolyte of dilute sulphuric acid. The sulphuric acid combines with the lead and the lead oxide to produce lead sulphate and water, electrical energy being released during the process. The overall reaction is:

$$Pb + PbO_2 + 2H_2SO_4 \longleftrightarrow 2PbSO_4 + 2H_2O$$

The reactions on each electrode of the battery are shown in Figure 2.4. In the upper part of the diagram the battery is discharging. Both electrode reactions result in the formation of lead sulphate. The electrolyte gradually loses the sulphuric acid, and becomes more dilute.

When being charged, as in the lower half of Figure 2.4, the electrodes revert to lead and lead dioxide. The electrolyte also recovers its sulphuric acid, and the concentration rises.

The lead acid battery is the most commonly used rechargeable battery in anything but the smallest of systems. The main reasons for this are that the main constituents (lead, sulphuric acid, a plastic container) are not expensive, that it performs reliably, and that it has a comparatively high voltage of about 2 V per cell. The overall characteristics of the battery are given in Table 2.1.

One of the most notable features of the lead acid battery is its extremely low internal resistance (see Section 2.2.1 and Figure 2.1). This means that the fall in voltage as current is drawn is remarkably small, probably smaller than for any of the candidate vehicle batteries. The figure given in Table 2.1 below is for a single cell, of nominal capacity 1.0 amphours. The capacity of a cell is approximately proportional to the area of the plates, and the internal resistance is approximately inversely proportional to the plate area. The result is that the internal resistance is, to a good approximation, inversely proportional to the capacity. The figure given in Table 2.1 of 0.022 Ω per cell is a rule of thumb figure taken from a range of good quality traction batteries. A good estimate of the internal resistance of a lead acid battery is thus:

$$R = \text{No. of cells} \times \frac{0.022}{C_{10}} \text{ Ohms} \qquad (2.3)$$

Batteries

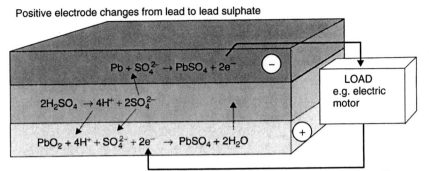

Reactions during the **discharge** of the lead acid battery.
Note that the electrolyte loses suphuric acid and gains water.

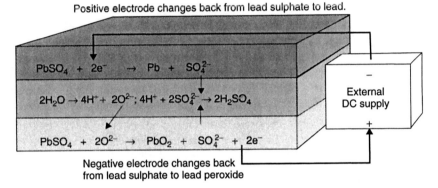

Reaction during the **charging** of the lead acid battery.
Note that the electrolyte suphuric acid concentration increases.

Figure 2.4 The reactions during the charge and discharge of the lead acid battery

Table 2.1 Nominal battery parameters for lead acid batteries

Specific energy	20–35 Wh.kg^{-1} depending on usage
Energy density	54–95 Wh.L^{-1}
Specific power	~250 W.kg^{-1} before efficiency falls very greatly
Nominal cell voltage	2 V
Amphour efficiency	~80%, varies with rate of discharge & temp.
Internal resistance	Extremely low, ~0.022 Ω per cell for 1 Amphour cell
Commercially available	Readily available from several manufacturers
Operating temperature	Ambient, poor performance in extreme cold
Self-discharge	~2% per day, but see text below
Number of life cycles	Up to 800 to 80% capacity
Recharge time	8 h (but 90% recharge in 1 h possible)

The number of cells is the nominal battery voltage divided by 2.0, six in the case of a 12 V battery. C_{10} is the Amphour capacity at the 10 hour rate.

2.3.2 Special characteristics of lead acid batteries

Unfortunately the lead acid battery reactions shown in Figure 2.4 are not the only ones that occur. The lead and lead dioxide are not stable in sulphuric acid, and decompose, albeit very slowly, with the reactions:

$$\text{At the positive electrode } 2PbO_2 + 2H_2SO_4 \longrightarrow 2PbSO_4 + 2H_2O + O_2 \quad (2.4)$$

$$\text{and at the negative } Pb + H_2SO_4 \longrightarrow PbSO_4 + H_2 \quad (2.5)$$

This results in the self-discharge of the battery. The rate at which these reactions occur depends on the temperature of the cell: faster if hotter. It also depends on other factors, such as the purity of the components (hence quality) and the precise alloys used to make up the electrode supports.

These unwanted reactions, that also produce hydrogen and oxygen gas, also occur while the battery is discharging. In fact they occur faster if the battery is discharged faster, due to lower voltage, higher temperature, and higher electrode activity. This results in the 'lost charge' effect that occurs when a battery is discharged more quickly, and which was noted in Section 2.2.2 above. It is a further unfortunate fact that these discharge reactions will not occur at exactly the same rate in all the cells, and thus some cells will become more discharged than others. This has very important consequences for the way batteries are charged, as explained below. But, in brief, it means that some cells will have to tolerate being 'over-charged' to make sure all the cells become charged.

The reactions that occur in the lead acid battery when it is being over-charged are shown in Figure 2.5. These gassing reactions occur when there is no more lead sulphate on the electrodes to give up or accept the electrons. They thus occur when the battery is fully or nearly fully charged.

We have noted that the charging and discharging reactions (as in Figure 2.4) involve changing the concentration of the electrolyte of the cells. The change in concentration of the reactants means that there is a small change in the voltage produced by the cell as it discharges. This decline in voltage is illustrated in Figure 2.6. For the modern sealed batteries the change is linear to quite a good approximation. It should be noted that this battery voltage *cannot* normally be used to give an indication of the state of charge of the battery. Is not normally possible to measure this open circuit voltage when the battery is in use, and in any case it is also greatly affected by temperature, and so a chance measurement of the battery voltage is likely to be strongly affected by other factors.

A notable feature of the overcharge reactions of Figure 2.5 and the self-discharge reactions of equations (2.4) and (2.5), is that water is lost and turned into hydrogen and oxygen. In older battery designs this gas was vented out and lost, and the electrolyte had to be topped up from time to time with water. In modern sealed batteries this is not necessary or even possible. The gases are trapped in the battery, and allowed to recombine (which happens at a reasonable rate spontaneously) to reform as water. Clearly there is

Batteries

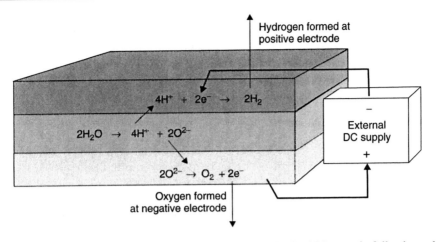

Figure 2.5 The gassing reactions that occur when the lead acid battery is fully charged

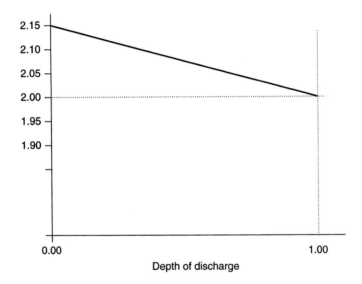

Figure 2.6 Graph showing how the open circuit voltage of a sealed lead acid battery changes with state of charge

a limit to the rate at which this can happen, and steps must be taken to make sure gas is not produced too rapidly. This is dealt with in the sections that follow.

Manufacturers of lead acid batteries can supply them in a wide range of heights, widths and lengths, so that for a given required volume they can be fairly accommodating. However, a problem with the wide use of lead acid batteries is that different designs are made for different applications, and it is essential to use the correct type. The type of battery used for the conventional car, the so-called starting, lighting and ignition (SLI) battery, is totally unsuitable for electric vehicle applications. Other lead acid batteries are designed for occasional use in emergency lighting and alarms; these are also totally unsuitable. The difference in manufacture is dealt with by authors such as Vincent and Scrosati (1998). It is only batteries of the 'traction' or 'deep cycling' type that are suitable here. This is the most expensive type of lead acid battery.

2.3.3 Battery life and maintenance

We have seen that gassing reactions occur within the lead acid battery, leading to loss of electrolyte. Traditional acid batteries require topping up with distilled water from time to time, but modern vehicle lead acid batteries are sealed to prevent electrode loss. In addition the electrolyte is a gel, rather than liquid. This means that maintenance of the electrolyte is no longer needed. However, the sealing of the battery is not total; there is a valve which releases gas at a certain pressure, and if this happens the water loss will be permanent and irreplaceable. This feature is a safety requirement, and leads to the name valve regulated sealed lead acid battery (VRLA) for this modern type of battery. Such a build up of gas will result if the reactions of Figure 2.5, which occur on overcharge, proceed too fast. This will happen if the charging voltage is too high. Clearly this must not be allowed to happen, or the battery will be damaged. On the positive side, it means that such batteries are essentially maintenance-free.

However, this does not mean that the batteries will last forever. Even if there is no water loss, lead acid batteries are subject to many effects that shorten their life. One of the most well known is the process called sulphation. This occurs if the battery is left for a long period (i.e. two weeks or more) in a discharged state. The lead sulphate (see Figure 2.4) on the electrodes forms into larger crystals, which are harder to convert back into lead or lead dioxide, and which form an insulating layer over the surface of the electrodes. By slowly recharging the battery this can sometimes be partially reversed, but often it cannot.

Making sure the battery is always kept in a good state of charge can prevent the problem of sulphation, and this is explained in Section 2.8. Section 2.8 also explains the very important issue of charge equalisation − getting this wrong is a major cause of battery failure. Other problems cannot be prevented however much care is taken. Within the electrodes of the battery corrosion reactions take place, which increase the electrical resistance of the contacts between the active materials that the electrode supports. The active materials will gradually form into larger and larger crystals, which will reduce the surface area, reducing both the capacity of the battery and slowing down the rate of reaction. The effects of vibration and the continual change of size of the active materials during the charge/discharge cycles (see Figure 2.4) will gradually dislodge them. As a result they will not make such good electrical contact with their support, and some will even fall off and become completely detached.

All these problems mean that the life of the lead acid battery is limited to around 700 cycles, though this strongly depends on the depth of the cycles. Experience with industrial trucks (fork-lifts, luggage carriers at railway stations, etc.) suggests that service lives of 1200–1500 cycles are possible, over 7–8 years. Fleet experience with electric cars indicates a life of about 5 years or 700 cycles. The shorter life of the road vehicles is the result of much greater battery load, the battery typically being discharged in about two hours, as opposed to the 7–8 hours for industrial trucks (Bosch 2000).

2.3.4 Battery charging

Charging a lead acid battery is a complex procedure and, as with any battery, if carried out incorrectly it will quickly ruin the battery and decrease its life. As we have seen, the charging must not be carried out at too high a voltage, or water loss results.

There are differing views on the best way of charging lead acid batteries and it is essential that, once a battery is chosen, the manufacturer's advice is sought.

The most commonly used technique for lead acid batteries is called multiple step charging. In this method the battery is charged until the cell voltage is raised to a predetermined level. The current is then switched off and the cell voltage is allowed to decay to another predetermined level and the current is then switched on again. A problem is that the predetermined voltages vary depending on the battery type, but also on the temperature. However, the lead acid battery is used in so many applications, that suitable good quality chargers are available from a wide range of suppliers.

An important point that applies to all battery types relates to the process of charge equalisation that must be done in all batteries at regular intervals if serious damage is not to result. It is especially important for lead acid batteries. This is fully explained below in Section 2.8, after all the main battery types have been described.

2.3.5 Summary of lead acid batteries

Lead acid batteries are well established commercially with good backup from industry. They are the cheapest rechargeable batteries per kilowatt-hour of charge, and will remain so for the foreseeable future. However, they have low specific energy and it is hard to see how a long-range vehicle can be designed using a lead acid battery. Lead acid will undoubtedly continue for some considerable time to be widely used for short-range vehicles. Lead acid batteries have a greater range of efficient specific powers than many other types (see Figure 2.3) and so they are very much in contention in hybrid electric vehicles, where only a limited amount of energy is stored, but it should be taken in and given out quickly.

2.4 Nickel-based batteries

2.4.1 Introduction

A range of commercial batteries using nickel in the positive electrode have been developed since Edison's work in the late 19th century. These batteries include nickel iron, nickel zinc, nickel cadmium and nickel metal hydride batteries. Two of these batteries are discussed below, the nickel metal hydride battery showing the most promise. The nickel zinc battery has a reasonable performance but it has a very limited life of 300 deep cycles and is not discussed further. The nickel/iron battery is also rarely used.

2.4.2 Nickel cadmium

The nickel cadmium battery was considered to be one of the main competitors to the lead acid battery for use in electric vehicles and these batteries have nearly twice the specific energy of lead acid batteries.

The NiCad battery uses nickel oxyhydroxide for the positive electrode and metallic cadmium for the negative electrode. Electric energy is obtained from the following reaction.

$$Cd + 2Ni\,O\,OH + 2H_2O \longleftrightarrow Cd(OH)_2 + 2Ni(OH)_2$$

The reactions at each electrode, which also help to explain 'where the electrons come from' and how the battery works, are shown in Figure 2.7. This battery makes an interesting comparison with the lead acid in that here the electrolyte becomes more concentrated as the cell discharges.

Nickel cadmium batteries have been widely used in many appliances, including use in electric vehicles. The NiCad battery has advantages of high specific power, a long life cycle (up to 2500 cycles), a wide range of operating temperatures from $-40°C$ to $+80°C$, a low self-discharge and good long term storage. This is because the battery is a very stable system, with equivalent reactions to the self-discharge of the lead acid battery (equations (2.4) and (2.5)) only taking place very slowly. The NiCad batteries can be purchased in a range of sizes and shapes, though they are not easy to obtain in the larger sizes required for electric vehicles, their main market being portable tools and electronic equipment. They are also very robust both mechanically and electrically and can be recharged within an hour and up to 60% capacity in 20 minutes.

On the negative side, the operating voltage of each cell is only about 1.2 V, so 10 cells are needed in each nominally 12 V battery, compared to 6 cells for lead acid. This partly explains the higher cost of this type of battery. A further problem is that the cost of cadmium is several times that of lead, and that this is not likely to change. Cadmium is also environmentally harmful and carcinogenic.

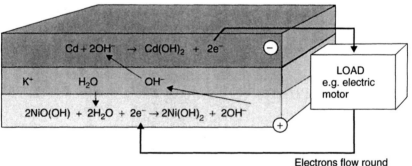

Reactions during the **discharge** of the NiCd battery.
Note that the electrolyte loses water, becoming more concentrated..

Electrons flow round the external circuit

Figure 2.7 Reactions during the discharge of a nickel cadmium cell. The reactions are reversed during charge

Batteries

Table 2.2 Nominal battery parameters for nickel cadmium batteries

Specific energy	40–55 Wh.kg^{-1} depending on current (see Figure 2.3)
Energy density	70–90 Wh.L^{-1} depending on current
Specific power	~125 W.kg^{-1} before becoming very inefficient
Nominal cell voltage	1.2 V
Amphour efficiency	Good
Internal resistance	Very low, ~0.06 Ω per cell for a 1 Amphour cell
Commercially available	Good in smaller sizes, difficult for larger batteries
Operating temperature	$-40°C$ to $+80°C$
Self-discharge	0.5% per day, very low
Number of life cycles	1200 to 80% capacity
Recharge time	1 h, rapid charge to 60% capacity 20 mins

The high cost of NiCad batteries, typically 3 times that of lead acid, is offset to an extent by its longer cycle life. Its charge efficiency decreases rapidly over 35°C but this is unlikely to affect its use in electric vehicles. It has been used successfully in cars such as electric versions of the Peugeot 106, the Citroen AX and the Renault Clio, as well as the Ford Th!nk® car shown in Figure 1.6. The overall characteristics of the battery are given in Table 2.2

As with lead acid batteries, NiCad batteries need to be properly charged. The points made in Section 2.8 below apply to this type of battery as well. However, because NiCad cells are less prone to self-discharge, the problem raised there is not so great as with lead acid. Normally the battery is charged at a constant current until its cell voltages reach a predetermined level, at which point the current is switched off. At this point the cell voltages decay to a lower predetermined voltage and the current is switched back on. This process is continued until the battery is recharged. A good proportion of the charge can be normally be replaced within 1 hour, but as explained in Section 2.8 the cell must be run at a fairly low current, with most of the cells being overcharged, for a longer time. Alternatively the battery can be recharged at a lower, constant current; this is a simpler system, but takes longer.

A clever feature of the NiCad battery is the way that it copes with overcharging. The cell is made so that there is a surplus of cadmium hydroxide in the negative electrode. This means that the positive electrode will always be fully charged first. A continuation of the charging current results in the generation of oxygen at the positive electrode via the reaction:

$$4OH^- \longrightarrow 2H_2O + O_2 + 4e^- \tag{2.6}$$

The resulting free oxygen diffuses to the negative electrode, where it reacts with the cadmium, producing cadmium hydroxide, using the water produced by reaction (2.6).

$$O_2 + 2Cd + 2H_2O \longrightarrow 2Cd(OH)_2 \tag{2.7}$$

As well as this reaction, the normal charging reaction will be taking place at this electrode, using the electrodes produced by reaction (2.4).

$$2Cd(OH)_2 + 4e^- \longrightarrow 2Cd + 4OH^- \tag{2.8}$$

Comparing reactions (2.7) and (2.8), we see that the rate of production of cadmium hydroxide is exactly equal to its rate of conversion back to cadmium. We thus have a perfectly sustainable system, with no net use of any material from the battery. The sum total of the reactions (2.6), (2.7) and (2.8) is no effect. This overcharging situation can thus continue indefinitely. For most NiCad batteries their size and design allows this to continue forever at the C/10 rate, i.e. at 10 A for a 100 Amphour battery. Of course this overcharging current represents a waste of energy, but it is not doing any harm to the battery, and is necessary in some cells while charging the battery in the final phase to equalise all the cells to fully charged.

It should be noted that although the internal resistance of the nickel cadmium battery is very low, it is not as low as for the lead acid battery. This results in a somewhat lower maximum economic specific power. The empirical, good 'first approximation' formula[2] for the internal resistance of a nickel cadmium battery is:

$$R = \text{No. of cells} \times \frac{0.06}{C_3} \text{ Ohms} \qquad (2.9)$$

Comparing this with equation (2.3) for the lead acid cell, it can be seen that there is a higher number (0.06 instead of 0.022). Also the number of cells will be greater, as has already been explained.

2.4.3 Nickel metal hydride batteries

The nickel metal hydride (NiMH) battery was introduced commercially in the last decade of the 20th century. It has a similar performance to the NiCad battery, the main difference being that in the NiMH battery the negative electrode uses hydrogen, absorbed in a metal hydride, which makes it free from cadmium, a considerable advantage.

An interesting feature of this battery type is that the negative electrode behaves exactly like a fuel cell, an energy source we consider more fully in the next chapter.

The reaction at the positive electrode is the same as for the nickel cadmium cell; the nickel oxyhydroxide becomes nickel hydroxide during discharge. At the negative electrode hydrogen is released from the metal to which it was temporarily attached, and reacts, producing water and electrons. The reactions at each electrode are shown in Figure 2.8.

The metals that are used to hold the hydrogen are alloys, whose formulation is usually proprietary. The principle of their operation is exactly the same as in the metal hydride hydrogen stores used in conjunction with fuel cells, and described in more detail in Section 5.3.5. The basic principle is a reversible reaction in which hydrogen is bonded to the metal, and then released as free hydrogen when required. For this to work the cell must be sealed, as an important driver in the absorption/desorption process is the pressure of the hydrogen gas, which is maintained at a fairly constant value. A further important point about the sealing is that the hydrogen-absorbing alloys will be damaged if air is allowed into the cell. This is because they will react with the air, and other molecules will occupy the sites used to store the hydrogen.

---

[2] The factor 0.06 in this formula is based on measurements from a small sample of good quality NiCad traction batteries.

Batteries

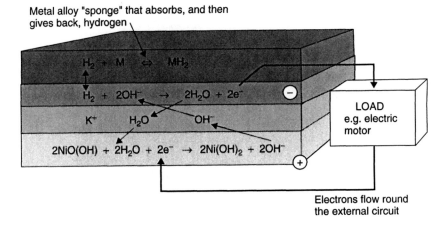

Figure 2.8 The reactions during the discharge of the nickel metal hydride cell. When charged the reactions are reversed. Note that when both discharging and charging water is created at exactly the same rate at which it is used, and that therefore the electrolyte does not change with state of charge

The overall chemical reaction for the NiMH battery is written as:

$$MH + NiOOH \longleftrightarrow M + Ni(OH)_2$$

In terms of energy density and power density the metal hydride cell is somewhat better than the NiCad battery. Ni/MH batteries have a nominal specific energy of about 65 Wh.kg^{-1} and a nominal energy density of 150 Wh.L^{-1} and a maximum specific power of about 200 W.kg^{-1}. Table 2.3 gives this and other information about this class of battery. In most respects its performance is similar to, or a little better than that for the nickel cadmium cell. The nominal cell voltage is 1.2 V.

One area where the NiMH is better than the NiCad is that it is possible to charge the battery somewhat faster. Indeed, it can be charged so fast that cooling becomes necessary.

Table 2.3 Nominal battery parameters for nickel metal hydride batteries

Specific energy	~65 Wh.kg^{-1} depending on power
Energy density	~150 Wh.L^{-1}
Specific power	200 W.kg^{-1}
Nominal cell voltage	1.2 V
Amphour efficiency	Quite good
Internal resistance	Very low, ~0.06 Ω per cell for a 1 Amphour cell
Commercially available	A good range of small cells, traction batteries difficult to obtain
Operating temperature	Ambient
Self-discharge	Poor, up to 5% per day
Number of life cycles	~1000 to 80% discharge
Recharge time	1 h, rapid charge to 60% capacity 20 mins

Figure 2.9 A commercial NiMH battery, with integral cooling fans

As well as heat energy being created by the normal internal resistance of the battery, the reaction in which hydrogen is bonded to the metal adjacent to the negative electrode is quite strongly exothermic. Unless the vehicle is a cycle or scooter, with a small battery, a cooling system is an important feature of NiMH battery systems. They are available commercially in small sizes, but larger batteries suitable for electric vehicles are beginning to appear. An example of a commercial NiMH battery is shown in Figure 2.9. Notice that the battery has cooling fans fitted as an integral part of the battery casing, for the reason explained above.

The NiMH battery has slightly higher energy storage capacity than NiCad systems, and is also a little more costly. There is one area where its performance is notably worse than that for NiCad batteries, and that its self-discharge properties. Hydrogen molecules are very small, and they can reasonably easily diffuse through the electrolyte to the positive electrode, where it will react:

$$\tfrac{1}{2}H_2 + Ni\,O\,OH \longrightarrow Ni(OH)_2$$

This effectively discharges the cell; hydrogen is lost from the negative and nickel hydroxide is formed at the positive. The result is that this battery is subject to quite rapid self-discharge.

An interesting feature of the cell, which can be seen by reference to Figure 2.8, is that the composition of the electrolyte does not change during charge or discharge; water and OH^- ions are created and used at exactly the same rate. The result is that the internal resistance and open circuit voltage of the cell are much more constant during discharge than with either lead acid or NiCad batteries. Being backed by a metal layer, the internal resistance is also a little lower, but it is not greatly different.

The charging regime is similar to that of the NiCad battery, the current being switched on and off to keep the cell voltage between an upper and a lower limit. Like NiCad batteries the NiMH battery can be charged within 1 hour. Most cells can cope with an overcharge current of about 0.1 C, like the NiCad cell. As will be explained in Section 2.8, overcharging is necessary in a battery to make sure each and every cell is fully charged.

Of all the new battery systems NiMH is considered to be one of the most advanced and has been used in a range of vehicles including the Toyota Prius, which has been by far the most successful electric hybrid to date. The market volume of NiMH batteries is still small, but with quantity production the price will drop. This battery is considered to be one of the most promising for the future.

2.5 Sodium-based Batteries

2.5.1 Introduction

In the 1980s a range of batteries which use a liquid sodium negative electrode were developed. These batteries differ from other batteries in so much as they run at high temperatures. They also have the interesting features of using one or more liquid electrodes in the form of molten sodium and using a solid ceramic electrolyte. Because of the need to operate at high temperatures they are only practical for large systems, such as electric cars; they are not suitable for scooters and cycles. They are rather more exotic than other types, as they will never be used in mobile phones or laptop computers, unlike the other types of battery that we will consider in this chapter. This limitation on their market has rather impeded their commercial development.

2.5.2 Sodium sulphur batteries

The development of these batteries started in the 1970s, and they run at temperatures between 300°C and 350°C. In order to keep the heat in the battery, the cells are enclosed in an evacuated case. The basic sodium sulphur cells, have a high specific energy, six times that of lead acid cells but in experimental batteries the mass of the enclosure typically halves this potential improvement.

The negative electrode in the cells consists of molten sodium, and the positive electrode consists of molten sulphur polysulphides. The electrolyte is a solid beta alumina ceramic, which conducts the sodium ions and also separates the two electrodes. The actual cells are kept fairly small and they are joined together and placed in an evacuated chamber to cut down heat losses. The design of the container needs careful thought as it can double the mass of the battery. Before the batteries can be used they have to be heated slowly to their working temperature. When in use the cells are essentially self-heating due to the electrical current passing through the battery internal resistance. When not in use for more than a day the battery interior has to be kept hot by the use of electrical heaters. The electrical energy is obtained by combining sodium with sulphur to form sodium sulphide.

The basic chemical formula for the reaction is

$$2Na + xS \longleftrightarrow Na_2S_x$$

Table 2.4 Nominal battery parameters for sodium sulphur batteries. As always, the performance figures depend on usage

Specific energy	100 Wh.kg^{-1} (Potentially 200 Wh.kg^{-1})
Energy density	150 Wh.L^{-1}
Specific power	200 W.kg^{-1}
Nominal cell voltage	2 V
Amphour efficiency	Very good
Internal resistance	Broadly similar to NiCad
Commercially available	Not on the market at all
Operating temperature	300–350°C
Self-discharge	Quite low, but when not in use energy must be supplied to keep the battery warm
Number of life cycles	~1000 to 80% capacity
Recharge time	8 h

The overall characteristics of the battery are given in Table 2.4. Because of the need for good thermal insulation small batteries are impractical. The battery heating and cooling needs careful design and management. Although the sodium sulphur battery has considerable promise, worries about the safety of two reactive materials separated by a brittle ceramic tube have largely resulted in the batteries not appearing on the commercial market. These fears were boosted by spontaneous fires involving test vehicles during trials.

2.5.3 Sodium metal chloride (Zebra) batteries

The sodium metal chloride or Zebra[3] battery is in many ways similar to the sodium sulphur battery and has many of this battery's advantages. However, with this system most (and some would say all) of the safety worries associated with the sodium sulphur battery have been overcome. The principle reason for the greater safety of the Zebra cells is the use of the solid positive electrolyte which is separated from the molten sodium metal by both solid and liquid electrolytes. It is certainly the case that prototype Zebra batteries have passed qualification tests for Europe, including rigorous tests such as crashing the cell at 50 kph into a steel pole (Vincent and Scrosati 1997, p. 272). This battery has considerable promise and it can be obtained commercially.

The Zebra cell uses solid nickel chloride for the positive electrode and molten sodium for the negative electrode. Two electrodes are used, a beta ceramic electrode surrounding the sodium and a secondary electrolyte, sodium-aluminium chloride, is used in the positive electrode chamber. Chlorine ions are the mobile ion in the electrolyte. The electrical energy on discharge is obtained by combining sodium with nickel chloride to give nickel and sodium chloride. The overall chemical reaction which takes place in the zebra battery is:

$$2Na + NiCl_2 \longleftrightarrow Ni + 2NaCl$$

---

[3] Zebra is an acronym for Zero Emissions Battery Research Association. However it has now rather lost this connection, and is used as a name for this type of battery.

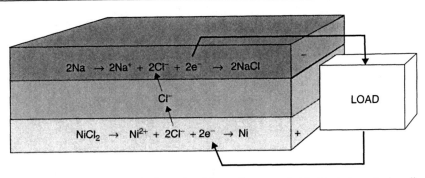

Figure 2.10 The reactions at each electrode of the sodium metal chloride battery during discharge

Figure 2.10 shows the reactions at each electrode during the middle and early part of the discharge of the cell. This reaction produces an open circuit voltage of about 2.5 V per cell. In the later stages of the discharge the reactions become more complex, involving aluminium ions from the electrolyte, and resulting in a lower voltage. Indeed an unfortunate feature of this type of cell is the way that the cell voltage falls during discharge, from about 2.5 V down to around 1.6 V. The internal resistance of the cell also increases, further affecting the output voltage. Nevertheless, as can be seen from the data in Table 2.5, the specific energy is very high, even with these effects.

A major problem with the Zebra battery is that it needs to operate at a temperature of about 320°C, similar to the sodium sulphur. Heat insulation is maintained by the use of a double skinned stainless steel box, with 2–3 cm of insulation between the two skins. All the air is removed from the insulation, and the vacuum is maintained for several years. Nevertheless, unless it is for a very short period, a few hours, these batteries need to be kept connected to a mains supply when not in use. This is to keep the battery hot, and is a major limitation to their application. As an example, the battery shown in Figure 2.11, which fits neatly under the seat of a battery electric car, holds an impressive 17.8 kWh of energy. However, when not in use it consumes about 100 W

Table 2.5 Nominal battery parameters for sodium metal chloride (Zebra) batteries

Specific energy	100 Wh.kg^{-1}
Energy density	150 Wh.L^{-1}
Specific power	150 W.kg^{-1}
Nominal cell voltage	~2 V average (2.5 V when fully charged)
Amphour efficiency	Very high
Internal resistance	Very low, but higher at low levels of charge
Commercially available	Available commercially, but very few suppliers
Operating temperature	300–350°C
Self-discharge	When not in use energy must be continually used to keep the battery up to temperature, corresponding to a self-discharge of about 10% per day
Number of life cycles	>1000
Recharge time	8 h

Figure 2.11 A commercial Zebra battery fitted neatly under the seat of an experimental battery electric vehicle by MES-DEA. The battery stores about 18 kWh of electrical energy

of power keeping up to temperature. So, in a 24 hour period the heating will require $0.1 \times 24 = 2.4$ kWh of energy, corresponding to about 13% of the stored energy. In energy terms, this corresponds to the self-discharge of other types of battery, and is quite a high figure.

Zebra batteries can be allowed to cool, but if this happens they must be reheated slowly and steadily, a process typically taking about 24 hours.

They are available as tried and tested units with well-established performance criteria, though only in a very limited range of size. An example is shown in Figure 2.11.

The overall characteristics of the battery are given in Table 2.5. These are taken from the 17.8 kWh (~280 V, 64 Ah, 180 kg, 32 kW peak power) unit manufactured by MES-DEA of Switzerland.

2.6 Lithium Batteries

2.6.1 Introduction

Since the late 1980s rechargeable lithium cells have come onto the market. They offer greatly increased energy density in comparison with other rechargeable batteries, though at greatly increased cost. It is a well-established feature of the most expensive laptop computers and mobile phones that lithium rechargeable batteries are specified, rather than the lower cost NiCad or NiHM cells that we have been considering earlier.

2.6.2 The lithium polymer battery

The lithium polymer battery uses lithium metal for the negative electrode and a transition metal intercalation oxide for the positive. In the resulting chemical reaction the lithium combines with the metal oxide to form a lithium metal oxide and release energy. When the battery is recharged the chemical reaction is reversed. The lithium is thus both a reactant and the mobile ion that moves through the electrolyte. The overall chemical reaction is:

$$xLi + M_yO_z \longleftrightarrow Li_xM_yO_z$$

The solid lithium negative electrode has been a cause of problems with this type of cell; there are safety difficulties and sometimes a decrease in performance due to passivation. Thus they have been largely superseded by the lithium ion battery.

2.6.3 The lithium ion battery

The lithium ion battery was introduced in the early 1990s and it uses a lithiated transition metal intercalation oxide for the positive electrode and lithiated carbon for the negative electrode. The electrolyte is either a liquid organic solution or a solid polymer.

Electrical energy is obtained from the combination of the lithium carbon and the lithium metal oxide to form carbon and lithium metal oxide. The overall chemical reaction for the battery is:

$$C_6Li_x + M_yO_z \longleftrightarrow 6C + Li_xM_yO_z$$

The essential features of the battery are shown in Table 2.6. An important point about lithium ion batteries is that accurate control of voltage is needed when charging lithium cells. If it is slightly too high it can damage the battery, and if too low the battery will be insufficiently charged. Suitable commercial chargers are being developed along with the battery.

Table 2.6 Nominal battery parameters for lithium ion batteries. These figures are based on the Sony lithium ion batteries as in 1998. Higher energy and power figures may be quoted, but these tend to be data for single cells or very small batteries, where packaging can be much more flimsy, and no provision need be made for cooling

Specific energy	90 Wh.kg^{-1}
Energy density	153 Wh.L^{-1}
Specific power	300 W.kg^{-1}
Nominal cell voltage	3.5 V
Amphour efficiency	Very good
Internal resistance	Very low
Commercially available	Only in very small cells not suitable for electric vehicles
Operating temperature	Ambient
Self-discharge	Very low, ~10% per month
Number of life cycles	>1000
Recharge time	2–3 h

The lithium ion battery has a considerable weight advantage over other battery systems, and this makes it a highly attractive candidate for future electric vehicle. The specific energy, for example, is about three times that of lead acid batteries, and this could give a car with a very reasonable range. However, large batteries are currently prohibitively expensive, and only when a commercial company has set up a production line which can produce lower-cost lithium ion batteries will their potential be fully realised. A few electric vehicles have been produced using lithium-based batteries, but they have been 'concept' type vehicles for demonstration purposes, to show what can be done. A notable example was an electric version of the Ford Ka produced in 2001 (Schmitz and Busch 2001).

2.7 Metal Air Batteries

2.7.1 Introduction

The metal air batteries represent an entirely different development, in the sense that the batteries cannot be recharged simply by reversing the current. Instead the spent metal electrodes must be replaced by new ones. The metal electrodes can thus be considered as a kind of fuel. The spent fuel is then sent to a reprocessing plant where it will be turned into new 'fuel'. The battery electrolyte will also normally need to be replaced.

This is not a dissimilar concept to conventional IC engine vehicles, where they stop periodically to refuel, with the added advantage that the vehicle will have the main attributes of an electrical vehicle: quietness and zero emissions. As such it may appeal to motorists who are used to refuelling their vehicles and may be slow to adapt to change.

2.7.2 The aluminium air battery

The basic chemical reaction of the aluminium air battery is essentially simple. Aluminium is combined with oxygen from the air and water to form aluminium hydroxide, releasing electrical energy in the process. The reaction is irreversible. The overall chemical reaction is:

$$4Al + 3O_2 + 6H_2O \longrightarrow 4Al(OH)_3$$

The aluminium forms the negative electrode of the cell, and it typically starts as a plate about 1 cm thick. As the reaction proceeds the electrode becomes smaller and smaller. The positive electrode is typically a porous structure, consisting of a metal mesh onto which is pressed a layer of catalysed carbon. A thin layer of PTFE gives it the necessary porosity to let the oxygen in, but prevent the liquid electrolyte getting out. The electrolyte is an alkaline solution, usually potassium hydroxide.

The battery is recharged by replacing the used negative electrodes. The electrolyte will normally also be replenished, as it will be contaminated with the aluminium hydroxide.

The essential characteristics of the aluminium air battery are shown in Table 2.7 The big drawback of the aluminium air battery is its extremely low specific power. For example a 100 kg battery would only give 1000 watts, which is clearly insufficient to power a

Batteries

Table 2.7 Nominal battery parameters for aluminium air batteries

Specific energy	225 Wh.kg^{-1}
Energy density	195 Wh.L^{-1}
Specific power	10 W.kg^{-1}
Nominal cell voltage	1.4 V
Amphour efficiency	N/A
Internal resistance	Rather high, hence low power
Commercially available	Stationary systems only available
Operating temperature	Ambient
Self-discharge	Very high (>10% per day) normally, but the electrolyte can be pumped out, which makes it very low
Number of life cycles	1000 or more
Recharge time	10 min, while the fuel is replaced

road vehicle. To get a power output of 20 kW, 2 tonnes of battery would be needed. The battery, on its own, is therefore not likely to be useful for most road vehicles.

2.7.3 The zinc air battery

The zinc air battery is similar in many ways to the aluminium air battery but it has a much better overall performance, particularly with regard to specific power which is nearly ten times that of the aluminium air battery, making it suitable for use in road vehicles. The structure is similar, with a porous positive electrode at which oxygen reacts with the electrolyte. The electrolyte is a liquid alkaline solution. The negative electrode is solid zinc.

The energy from the battery is obtained by combining zinc with the oxygen in the air and forming zinc oxide. Alternatively, depending on the state of the electrodes and electrolyte, zinc hydroxide may be formed, as for the aluminium-air cell. The process is normally irreversible.

The general characteristics of the battery are shown in Table 2.8. A few manufacturers have claimed to produce electrically rechargeable zinc-air batteries, but the number of

Table 2.8 Nominal battery parameters for zinc air batteries

Specific energy	230 Wh.kg^{-1}
Energy density	270 Wh.L^{-1}
Specific power	105 W.kg^{-1}
Nominal cell voltage	1.2 V
Amphour efficiency	Not applicable
Internal resistance	medium
Commercially available	A very few suppliers
Operating temperature	Ambient
Self-discharge	High, as electrolyte is left in cell
Number of life cycles	>2000
Recharge time	10 min, while the fuel is replaced

cycles is usually quite small. The more normal way of recharging is as for the aluminium-air cell, which is by replacing the negative electrodes. The electrolyte, containing the zinc oxide, is also replaced. In principle this could be taken back to a central plant, and the zinc recovered, but the infrastructure for doing this would be rather inconvenient.

Small zinc air batteries have been available for many years, and their very high energy density makes them useful in applications such as hearing aids. These devices are usually 'on' virtually all the time, and so the self-discharge is not so much of a problem. Large batteries, with replaceable negative electrodes, are only available with great difficulty, but this is changing and they show considerable potential for the future. Use of a replaceable fuel has considerable advantages as it avoids the use of recharging points, lorries delivering fuel can simply take the spent fuel back to the reprocessing plant from where they got it in the first place. The high specific energy will also allow reasonable journey times between stops.

2.8 Battery Charging

2.8.1 Battery chargers

The issue of charging batteries is of the utmost importance for maintaining batteries in good order and preventing premature failure. We have already seen, for example, how leaving a lead acid battery in a low state of charge can cause permanent damage through the process of sulphation. However, charging them improperly can also very easily damage batteries.

Charging a modern vehicle battery is not a simple matter of providing a constant voltage or current through the battery, but requires very careful control of current and voltage. The best approach for the designer is to buy commercial charging equipment from the battery manufacturer or another reputed battery charger manufacturer. When the vehicle is to be charged in different places where correct charging equipment is not available, the option of a modern light onboard charger should be considered.

Except in the case of photoelectric panels, the energy for recharging a battery will nearly always come from an alternating current (AC) source such as the mains. This will need to be rectified to direct current (DC) for charging the battery. The rectified DC must have very little ripple, it must be very well 'smoothed'. This is because at the times when the variation of the DC voltage goes below the battery voltage, no charging will take place, and at the 'high point' of the ripple it is possible that the voltage could be high enough to damage the battery. The higher the DC current, the harder it is for rectifiers to produce a smooth DC output, which means that the rectifying and smoothing circuits of battery chargers are often quite expensive, especially for high current chargers. For example, the battery charger for the important development vehicle, the General Motors EV1, cost about $2000 in 1996 (Shnayerson 1996).

One important issue relating to battery chargers is the provision of facilities for charging vehicles in public places such as car parks. Some cities in Europe, especially (for example) La Rochelle in France, and several in California in the USA, provide such units. A major problem is that of standardisation, making sure that all electric vehicles can safely connect to all such units. Recently the Californian Air Resources Board, which regulates such matters there, has produced guidelines, which are described elsewhere (Sweigert *et al.* 2001). This paper also gives a good outline of the different ways in which these car-to-charger connections can be made.

However, the great majority of electric vehicles, such as bicycles, mobility aids, delivery vehicles and the like, will always use one charger, which will be designed specifically for the battery on that vehicle. On hybrid electric vehicles too, the charger is the alternator on the engine, and the charging will be controlled by the vehicle's energy management system. However, whatever charging method is used, with whatever type of battery, the importance of 'charge equalisation' in batteries must be understood. This is explained in the following section.

2.8.2 Charge equalisation

An important point that applies to all battery types relates to the process of charge equalisation that must be done in all batteries at regular intervals if serious damage is not to result.

A problem with all batteries is that when current is drawn not all the individual cells in the battery lose the same amount of charge. Since a battery is a collection of cells connected in series, this may at first seem wrong; after all, exactly the same current flows through them all. However, it does not occur because of different currents (the electric current is indeed the same) it occurs because the *self-discharge effects* we have noted (e.g. equations (2.4) and (2.5) in the case of lead acid batteries) take place at different rates in different cells. This is because of manufacturing variations, and also because of changes in temperature; the cells in a battery will not all be at exactly the same temperature.

The result is that if nominally 50% of the charge is taken from a battery, then some cells will have lost only a little more than this, say 52%, while some may have lost considerably more, say 60%. If the battery is recharged with enough for the good cell, then the cells more prone to self-discharge will not be fully re-charged. The effect of doing this repeatedly is shown in Table 2.9.

Cell A cycles between about 20% and 80% charged, which is perfectly satisfactory. However, Cell B sinks lower and lower, and eventually fails after a fairly small number

Table 2.9 Showing the state of charge of two different cells in a battery. Cell A is a good quality cell, with low self-discharge. Cell B has a higher self-discharge, perhaps because of slight manufacturing faults, perhaps because it is warmer. The cells are discharged and charged a number of times

State of charge of cell A	State of charge of cell B	Event
100%	100%	Fully charged
48%	40%	50% discharge
98%	90%	50% charge replaced
35%	19%	60% discharge
85%	69%	50% partial recharge
33%	9%	50% discharge
83%	59%	50% partial recharge
18%	Cannot supply it, battery flat!	60% discharge required to get home

of cycles.[4] If one cell in a battery goes completely flat like this, the battery voltage will fall sharply, because the cell is just a resistance lowering the voltage. If current is still drawn from the battery, that cell is almost certain to be severely damaged, as the effect of driving current through it when flat is to try and charge it the 'wrong way'. Because a battery is a series circuit, one damaged cell ruins the whole battery. *This effect is probably the major cause of premature battery failure.*

The way to prevent this is to fully charge the battery till each and every cell is fully charged (a process known as charge equalisation) at regular intervals. This will inevitably mean that some of the cells will run for perhaps several hours being overcharged. Once the majority of the cells have been charged up, current must continue to be put into the battery so that those cells that are more prone to self-discharge get fully charged up.

This is why it is important that a cell can cope with being overcharged. However, as we have seen in Sections 2.3 and 2.4, only a limited current is possible at overcharge, typically about $C/10$. For this reason the final process of bringing all the cells up to fully charged cannot be done quickly. This explains why it takes so much longer to fully charge a battery than to take it to nearly full. The last bit has to be done slowly. It also explains something of the complexity of a good battery charger, and why the battery charging process is usually considerably less than 100% charge efficient. Figure 2.12 shows this process, using an example not quite so extreme as the data in Table 2.9. Unlike in Table 2.9, the battery in Figure 2.12 is 'saved' by ensuring that charge equalisation takes place before any cells become completely exhausted of charge.

So far we have taken to process of charge equalisation to be equalising all the calls to full. However, in theory it is possible to equalise the charge in all cells of the battery at any point in the process, by moving charge from one cell to the other, from the more charged to the less charged. This is practical in the case of the 'super-capacitors' considered in the next chapter, however it is not usually practical with batteries. The main reason is the difficulty of sensing the state of charge of a cell, whereas for a capacitor it is much easier, as the voltage is directly proportional to charge. However, in the case of lithium-based batteries charge equalisation by adding circuits to the battery system is more practical, and is used. Chou *et al.* (2001) give a good description of such a battery management system.

This issue of some cells slowly becoming more deeply discharged than others is very important in battery care. There are two particular cases where it is especially important.

Opportunistic charging: some users are able to put a small amount of charge back into a battery, for example when parked in a location by a charger for a short time. This is helpful, but the user MUST make certain that fairly frequently a full long charge is given to the battery to bring all cells up to 100% charged.

Hybrid electric vehicles: in these it is desirable to have the battery NOT fully charged normally, so that the battery can always absorb energy from regenerative braking. However, this must be done with caution, and the battery management system must periodically run the battery to fully charged to equalise all the cells to 100% charged.

---

[4] The very large difference in self-discharge of this example is somewhat unlikely. Nevertheless, the example illustrates what happens, though usually more slowly than the four cycles of Table 2.9.

Batteries

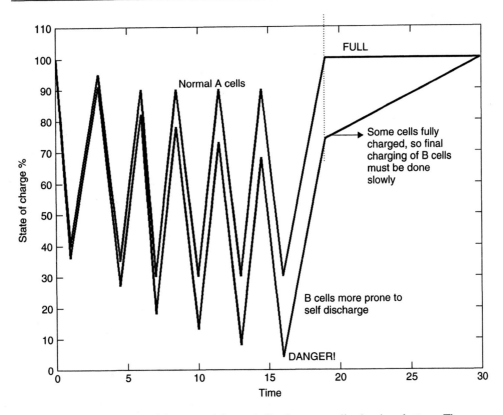

Figure 2.12 Diagram showing the need for periodic charge equalisation in a battery. The upper line (A) shows the state of charge of a normal cell working satisfactorily. The lower line (B) is for a cell more prone to self-discharge. Charge equalisation involves overcharging some of the cells while the others are brought up to full charge. This is occurring in the final 12 time units

The issues of battery charging mentioned here apply to all battery types. However, they are more important for cells with higher self discharge rates, such as the lead acid. The only batteries for which this is not of the utmost importance are the small single cells used in electronic products; however, they are not relevant here.

2.9 The Designer's Choice of Battery

2.9.1 Introduction

At first glance the designer's choice of battery may seem a rather overwhelming decision. In practice it is not that complicated, although choosing the correct size of battery may be. Firstly the designer needs to decide whether he/she is designing a vehicle which will use batteries that are currently available either commercially, or by arrangement with battery manufacturers for prototype use. Alternatively the designer may be designing a futuristic vehicle for a client or as an exercise, possibly as part of an undergraduate course. The designer will also need to decide on the specification and essential requirements of the vehicle. Is he, for example, designing the vehicle for speed, range, capital cost, running costs, overall costs, style, good handling, good aerodynamics, environmentally friendliness, etc.; is he they looking for an optimum design that takes many of these design parameters into consideration. Also, is he they considering a hybrid or non-hybrid vehicle?

2.9.2 Batteries which are currently available commercially

Of the batteries discussed in this chapter the ones which are now available commercially for use in electric vehicles include: lead acid, nickel cadmium, nickel metal hydride, sodium metal chloride (Zebra) and lithium ion. For comparative purposes these batteries are shown in Table 2.10

Of the batteries mentioned above lithium ion is prohibitively expensive, unless of course you are designing an electric racing car, with no expense spared, in which case this may be your chosen option. Leaving aside the racing car for the moment, this narrows the choice to lead acid, nickel metal hydride and sodium metal chloride.

For a long term study there is no substitute for making a mathematical model of the vehicle using information supplied later in the book and comparing the results using different batteries. However, for some vehicles the battery choice is fairly obvious and the mathematical model can simply be used to confirm vehicle size and overall performance.

For example, lead acid is cheap, and for uses not requiring large amounts of energy storage (e.g. for short range vehicles such as golf carts and wheel chairs, which can be charged overnight), there is no better choice of battery. Lead acid is widely used, has a long track record and has the lowest cost per kWh of storage capacity.

Nickel metal hydride is a good choice where range and performance are needed. It also can be recharged very quickly, and for uses where the vehicle can be charged frequently

Table 2.10 Comparison of commercially available batteries. The cost figure is an arbitrary unit for broad comparative purposes. All the other figures are also very much guidelines only; we have explained that all such performance figures depend on how the battery is used

Battery	Specific energy Wh.kg^{-1}	Energy density Wh.L^{-1}	Specific power W.kg^{-1}	Current cost
Lead acid	30	75	250	0.5
NiCad	50	80	150	1.5
NiMH	65	150	200	2.0
Zebra	100	150	150	2.0
Li-ion[5]	90	150	300	10
Zinc-air	230	270	105	?

[5] Much higher performance figures are sometimes quoted for this type of battery. However, these are nearly always for single cells or small batteries. By the time the necessary packaging and cooling equipment have been added, the figures come into the region of those shown here.

this could result in a smaller and cheaper battery unit than, for instance, if a lead acid battery were used. It would therefore come into its own for hybrid vehicles or vehicles such as a commuter bus or tram that stop frequently and could therefore be charged when stopped.

Sodium metal chloride (Zebra) batteries are not used in small sizes because the heat losses are proportionally large. The commercial battery, shown in Figure 2.11, for example, has 17.8 kWh of storage. The Zebra battery has many of the attributes of nickel metal hydride, but with even greater energy density. However, the fact that it needs to be kept hot is a major drawback to its use in IC engine/electric hybrids, as these are largely totally autonomous vehicles, which may be left unused for long periods, for example in an airport car park during a two week holiday.

Lithium ion can currently be used where high performance rather than cost is the main criterion.

2.10 Use of Batteries in Hybrid Vehicles

2.10.1 Introduction

There are many combinations of batteries, engines and mechanical flywheels which allow optimisation of electric vehicles. The best known is the combination of IC engine and rechargeable battery, but more than one type of battery can be used in combination, and the use of batteries and flywheels can have advantages.

2.10.2 Internal combustion/battery electric hybrids

Where IC engine efficiency is to be optimised by charging and supplying energy from the battery, clearly a battery which can be rapidly charged is desirable. This tends to emphasise batteries such as the nickel metal hydride, which is efficient and readily charged and discharged. Examples of this would be the Toyota Prius and the Honda Insight, both very successful hybrids that use nickel metal hydride batteries. A zinc air battery would be no use in this situation, as it cannot be electrically recharged.

This type of hybrid electric vehicle, IC engine with battery, is by the most common, and is likely to be the most important type of electric car in the near and even medium term. It seems that the majority of such vehicles currently use nickel metal hydride batteries, with a storage capacity typically between about 2 and 5 kWh. (Note that the energy stored in a normal car battery is between about 0.3 and 1.0 kWh.)

2.10.3 Battery/battery electric hybrids

Different batteries have different characteristics and they can sometimes be combined to give optimum results. For example, an aluminium air battery has a low specific power and cannot be recharged, but could be used in combination with a battery which recharges and discharges quickly and efficiently, such as the nickel metal hydride battery. The aluminium air battery could supply a base load that sends surplus electricity to the NiMH battery when the power is not required. The energy from the NiMH battery could then be supplied for accelerating in traffic or overtaking; it could also be used for accepting and resupplying electricity for regenerative braking.

2.10.4 Combinations using flywheels

Flywheels that drive a vehicle through a suitable gearbox can be engineered to store small amounts of energy quickly and efficiently and resupply it soon afterwards. They can be used with mechanisms such as a cone/ball gearbox. They can be usefully employed with batteries that could not do this. For example the zinc air battery cannot be recharged in location in the vehicle, and hence cannot be used for regenerative braking, but by combining this with a suitable flywheel a vehicle using a zinc air battery with regenerative braking could be designed

2.10.5 Complex hybrids

There is plenty of scope for originality from designers. Two or more batteries, an IC engine and a flywheel for example, may achieve the optimum combination. Alternatively a fuel cell could be combined with a battery and a flywheel.

2.11 Battery Modelling

2.11.1 The purpose of battery modelling

Modelling (or simulating) of engineering systems is always important and useful. It is done for different reasons. Sometimes models are constructed to understand the effect of changing the way something is made. For example, we could construct a battery model that would allow us to predict the effect of changing the thickness of the lead oxide layer of the negative electrodes of a sealed lead acid battery. Such models make extensive use of fundamental physics and chemistry, and the power of modern computers allows such models to be made with very good predictive powers.

Other types of model are constructed to accurately predict the behaviour of a particular make and model of battery in different circumstances. This model will then be used to predict the performance of a vehicle fitted with that type of battery. This sort of model relies more on careful analysis of real performance data than fundamental physics and chemistry.

In this book we will concern ourselves only with the latter type of performance modelling. However, all modelling of batteries is notoriously difficult and unreliable. The performance of a battery depends on reasonably easily measurable quantities such as its temperature, and performance characteristics such as voltage. However, it also depends on parameters far harder to specify precisely, such as age, and the way the battery has been used (or misused) in the past. Manufacturing tolerances and variation between the different cells within a battery can also have a big impact on performance.

The result of these problems is that all we can do here is give an introduction to the task of battery simulation and modelling.

2.11.2 Battery equivalent circuit

The first task in simulating the performance of a battery is to construct an equivalent circuit. This is a circuit made up of elements, and each element has precisely predictable behaviour.

We introduced such an equivalent circuit at the beginning of this chapter. Figure 2.1 is a very simple (but still highly useful) equivalent circuit for a battery. A limitation of this type of circuit is that it does not explain the dynamic behaviour of the battery at all. For example, if a load is connected to the battery the voltage will immediately change to a new (lower) value. In fact this is not true; rather, the voltage takes time to settle down to a new value.

Figure 2.13 shows a somewhat more refined equivalent circuit that simulates or models these dynamic effects quite well. We could carry on refining our circuit more and more to give an ever-closer prediction of performance. These issues are discussed in the literature, for example by Johnson *et al.* (2001).

The purpose of our battery simulations is to be able to predict the performance of electric vehicles, in terms of range, acceleration, speed and so on, a topic covered in reasonable depth in Chapter 7. In these simulations the speed of the vehicles changes fairly slowly, and the dynamic behaviour of the battery makes a difference that is small compared to the other approximations we have to make along the way. Therefore, in this introduction to battery simulation we will use the basic equivalent circuit of Figure 2.1.

Although the equivalent circuit of Figure 2.1 is simple, we do need to understand that the values of the circuit parameters (E and R) are not constant. The open circuit voltage of the battery E is the most important to establish first. This changes with the state of charge of the battery.

In the case of the sealed lead acid battery we have already seen that the open circuit voltage E is approximately proportional to the state of charge of the battery, as in Figure 2.6. This shows the voltage of one cell of a battery. If we propose a battery variable DoD, meaning the depth of discharge of a battery, which is zero when fully charged

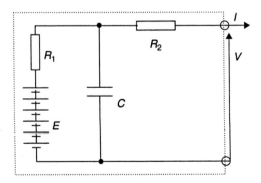

Figure 2.13 Example of a more refined equivalent circuit model of a battery. This models some of the dynamic behaviour of a battery

and 1.0 when empty, then the simple formula for the open circuit voltage is:

$$E = n \times (2.15 - DoD \times (2.15 - 2.00)) \tag{2.10}$$

where n is the number of cells in the battery. This formula gives reasonably good results for this type of battery, though a first improvement would be to include a term for the temperature, because this has a strong impact.

In the case of nickel-based batteries such a simple formula cannot be constructed. The voltage/state of charge curve is far from linear. Fortunately it now very easy to use mathematical software, such as MATLAB®, to find polynomial equations that give a very good fit to the results. One such, produced from experimental results from a NiCad traction battery is:

$$E = n \times \begin{pmatrix} -8.2816 DoD^7 + 23.5749 DoD^6 - 30 DoD^5 + 23.7053 DoD^4 \\ -12.5877 DoD^3 + 4.1315 DoD^2 - 0.8658 DoD + 1.37 \end{pmatrix} \tag{2.11}$$

The purpose of being able to simulate battery behaviour is to use the results to predict vehicle performance. In other words we wish to use the result in a larger simulation. This is best done in software such as MATLAB® or an EXCEL® spreadsheet. Which program is used depends on many factors, including issues such as what the user is used to. For the purposes of a book like this, MATLAB® is the most appropriate, since it is very widely used, and it is much easier than EXCEL® to explain what you have done and how to do it.

A useful feature of MATLAB® is the ability to create functions. Calculating the value of E is a very good example of where such a function should be used. The MATLAB® function for finding E for a lead acid battery is as follows:

```
function E_oc=open_circuit_voltage_LA(x,N)
% Find the open circuit voltage of a lead acid
% battery at any value of depth of discharge
% The depth of discharge value must be between
% 0 (fully charged) and 1.0 (flat).

if x<0
      error('Depth of discharge <0.');
end
if x > 1
   error('Depth of discharge >1')
end

% See equation >2.10 in text.

E_oc = (2.15 - ((2.15-2.00)*x)) * N;
```

The function for a NiCad battery is identical, except that the last line is replaced by a formula corresponding to equation (2.11).

Our very simple battery model of Figure 2.1 now has a means of finding E, at least for some battery types. The internal resistance also needs to be found. The value of R

is approximately constant for a battery, but it is affected by the state of charge and by temperature. It is also increased by misuse, and this is especially true of lead acid batteries. Simple first-order approximations for the internal resistance of lead acid and nickel-based batteries have been given in equations (2.3) and (2.9).

2.11.3 Modelling battery capacity

We have seen in Section 2.2.2 that the capacity of a battery is reduced if the current is drawn more quickly. Drawing 1 A for 10 hours does not take the same charge from a battery as running it at 10 A for 1 hour.

This phenomenon is particularly important for electric vehicles, as in this application the currents are generally higher, with the result that the capacity might be less than is expected. It is important to be able to predict the effect of current on capacity, both when designing vehicles, and when making instruments that measure the charge left in a battery: battery fuel gauges. Knowing the depth of discharge of a battery is also essential for finding the open circuit voltage using equations such as (2.10) and (2.11).

The best way to do this is using the Peukert model of battery behaviour. Although not very accurate at low currents, for higher currents it models battery behaviour well enough.

The starting point of this model is that there is a capacity, called the Peukert Capacity, which is constant, and is given by the equation:

$$C_p = I^k T \tag{2.12}$$

where k is a constant (typically about 1.2 for a lead acid battery) called the Peukert Coefficient. This equation assumes that the battery is discharged until it is flat, at a constant current I A, and that this will last T h. Note that the Peukert Capacity is equivalent to the normal Amphours capacity for a battery discharged at 1 A. In practice the Peukert Capacity is calculated as in the following example.

Suppose a battery has a nominal capacity of 40 Ah at the 5 h rate. This means that it has a capacity of 40 Ah if discharged at a current of:

$$I = \frac{40}{5} = 8 \text{ A} \tag{2.13}$$

If the Peukert Coefficient is 1.2, then the Peukert Capacity is:

$$C_p = 8^{1.2} \times 5 = 60.6 \text{ Ah} \tag{2.14}$$

We can now use equation (2.12) (rearranged) to find the time that the battery will last at any current I.

$$T = \frac{C_p}{I^k} \tag{2.15}$$

The accuracy of this Peukert model can be seen by considering the battery data shown in Figure 2.2. This is for a nominally 42 Ah battery (10 h rate), and shows how the capacity changes with discharge time. This solid line in Figure 2.14 shows the data of Figure 2.2 in

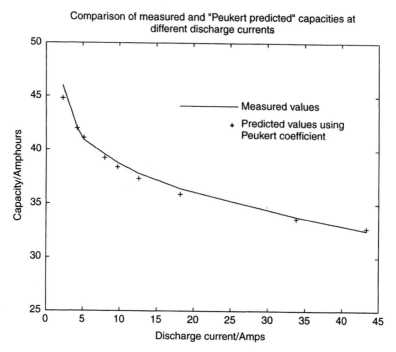

Figure 2.14 Showing how closely the Peukert model fits real battery data. In this case the data is from a nominally 42 V lead acid battery

a different form, i.e. it shows how the capacity declines with increasing discharge current. Using methods described below, the Peukert Coefficient for this battery has been found to be 1.107. From equation (2.12) we have:

$$C_p = 4.2^{1.107} \times 10 = 49 \text{ Ah}$$

Using this, and equation (2.15), we can calculate the capacity that the Peukert equation would give us for a range of currents. This has been done with the crosses in Figure 2.14. As can be seen, these are quite close to the graph of the measured real values.

The conclusion from equations (2.12) and (2.15) is that if a current I flows from a battery, then, from the point of view of the battery capacity, the current that appears to flow out of the battery is I^k A. Clearly, as long as I and k are greater than 1.0, then I^k will be larger than I.

We can use this in a real battery simulation, and we see how the voltage changes as the battery is discharged. This is done by doing a step-by-step simulation, calculating the charge removed at each step. This can be done quite well in EXCEL®, but for reasons explained earlier MATLAB® will be used here.

The time step between calculations we will call δt. If the current flowing is I A, then the apparent or effective charge removed from the battery is:

$$\delta t \times I^k \tag{2.16}$$

There is a problem of units here. If δt is in seconds, this will be have to be divided by 3600 to bring the units into Amphours. If CR_n is the total charge removed from the battery by the nth step of the simulation, then we can say that:

$$CR_{n+1} = CR_n + \frac{\delta t \times I^k}{3600} \text{ Ah} \qquad (2.17)$$

It is very important to keep in mind that this is the charge removed from the plates of the battery. It is not the total charge actually supplied by the battery to the vehicle's electrics. This figure, which we could call CS (charge supplied), is given by the formula:

$$CS_{n+1} = CS_n + \frac{\delta t \times I}{3600} \text{Ah} \qquad (2.18)$$

This formula will normally give a *lower* figure. As we saw in the earlier sections, this difference is caused by self-discharge reactions taking place *within* the battery.

The depth of discharge of a battery is the ratio of the charge removed to the original capacity. So, at the nth step of a step-by-step simulation we can say that:

$$DOD_n = \frac{CR_n}{C_p} \qquad (2.19)$$

Here C_p is the Peukert Capacity, as from equation (2.12). This value of depth of discharge can be used to find the open circuit voltage, which can then lead to the actual terminal voltage from the simple equation already given as Equation (2.1).

To simulate the discharge of a battery these equations are 'run through', with n going from 1, 2, 3, 4, etc., until the battery is discharged. This is reached when the depth of discharge is equal to 1.0, though it is more common to stop just before this, say when *DoD* is 0.99.

The script file below runs one such simulation for a NiCad battery.

```
% Simple battery constant current discharge experiment for
% a large 5 cell NiCad battery. The time step is set to 50
% seconds, which is sufficiently small for such a constant
% current experiment.

% We need to form some arrays for holding data. The array T
% is for time, which will run from 0 to 50000 seconds, in 50
% 50 second steps.
T=(0:50:50000);
% This corresponds to 1001 values. We form four more arrays,
% each also with 1001 elements, and all with initial values
% of zero. Dod(n) is used to store values of the depth of
% discharge, V(n) stores voltage values, CR(n) and CS(n)
% store values of the charge, in Amphours, removed from the
% battery and supplied by the battery.
CR=zeros(1,1001);     % Charged removed from electrodes,
                      % corrected using Peukert coefficient.
DoD=zeros(1,1001);    % Depth of discharge, start off fully
                      % charged.
```

```
V=zeros(1,1001); % Battery voltage at each time step
CS=zeros(1,1001);   % Charge supplied by the battery in Ah

% We now set some constants for the experiment
I = 30;        % Set discharge current to 30 Amps
NoCells=5; % 5 cell battery
Capacity=50;   % This is the normal 3 h rated capacity of the
               % battery
k=1.045;       % Peukert coefficient, not much greater than 1.
deltaT = 50;   % Take 10 second time steps, OK for con I.
% Calculated values
Rin = (0.06/Capacity)*NoCells; % Internal resistance, eq 2.9
PeuCap = ((Capacity/3)^k)*3;    % See equation 2.12
% Starting voltage set outside loop
V(1) = open_circuit_voltage_NC(0,NoCells) - I*Rin; % Equ 2.1

for n=2:1001
    CR(n) = CR(n-1) + ((I^k * deltaT)/3600); % Equation 2.17
    DoD(n) = CR(n)/PeuCap;    % Equation 2.19
    if DoD(n)>1
       DoD(n)=1;
    end

    V(n)=open_circuit_voltage_NC(DoD(n),NoCells) - I*Rin;
    % We will say that the battery is "dead" if the
    % depth of discharge exceeds 99%
    if DoD(n)>0.99
       V(n)=0;
    end

    % We now calculate the real amphours given out by the
    % battery. This uses the actual current, NOT Peukert
    % corrected.
    if V(n)>0
       CS(n)=CS(n-1)+ ((I*deltaT)/3600); % Equation 2.18
    else
       CS(n)=CS(n-1);
    end
end
%The bat. V could be plotted against t, but it is sometimes
% more useful to plot against Ah given out. This we do here.
plot(CS,V,'b.');
axis([0 55 3.5 7]);
XLABEL('Charge supplied/Amphours');
YLABEL('Battery voltage/Volts');
TITLE('Constant current discharge of a 50 Ah NiCad battery');
```

This script file runs the simulation at one unchanging current. Figure 2.15 shows the graphs of voltage for three different currents. The voltage is plotted against the actual charge supplied by the battery, as in equation (2.18). The power of this type of simulation can be seen by comparing Figure 2.15 with Figure 2.16, which is a copy of the similar data taken from measurements of the real battery.

Batteries

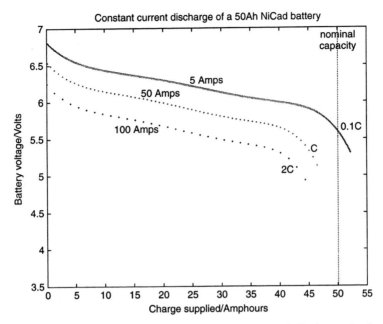

Figure 2.15 Showing the voltage of a 6 V NiCad traction battery as it discharges for three different currents. These are simulated results using the model described in the text

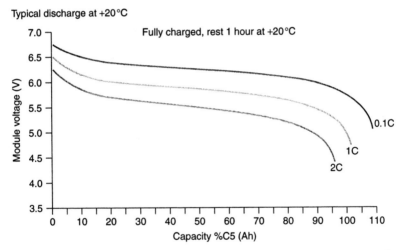

Figure 2.16 Results similar to those of Figure 2.15, but these are measurements from a real battery.

2.11.4 Simulation a battery at a set power

When making a vehicle go at a certain speed, then it is a certain *power* that will be required from the motor. This will then require a certain electrical power from the battery. It is thus useful to be able to simulate the operation of a battery at a certain set power, rather than current.

The first step is to find an equation for the current I from a battery when it is operating at a power P W. In general we know that:

$$P = V \times I$$

If we then combine this with the basic equation for the terminal voltage of a battery, which we have written as equation (2.1), we get:

$$P = V \times I = (E - IR) \times I = EI - RI^2$$

This is a quadratic equation for I. The normal useful solution[6] to this equation is:

$$I = \frac{E - \sqrt{E^2 - 4RP}}{2R} \qquad (2.20)$$

This equation allows us to easily use MATLAB® or similar mathematical software to simulate the constant power discharge of a battery. The MATLAB® script file below shows this done for a lead acid battery. The graph of voltage against time is shown in Figure 2.17.

```
% A constant P discharge experiment for a lead acid battery.
% The system has 10 batteries, each 12 V lead acid, 50 Ah.
% We use 10 s steps, as these are sufficiently small for
% a constant power discharge. We set up arrays to store the
% data.
T=(0:10:10000);    % Time goes up to 10,000 in 10 s steps.
                   % This is 1001 values.
CR=zeros(1,1001);  % Charge rem. from bat. Peukert corrected.
DoD=zeros(1,1001); % Depth of dis., start fully charged.
V=zeros(1,1001);   % Battery voltage, initially set to zero.
NoCells=60;        % 10 of 6 cell (12 Volt) batteries.
Capacity=50;       % 50 Ah batteries, 10 h rate capacity
k=1.12;            % Peukert coefficient
deltaT = 10;       % Take 10s steps, OK for constant power.
P = 5000;          % We will drain the battery at 5 kW.
% Calculated values
Rin = (0.022/Capacity)*NoCells;    % Internal re, Equ. 2.2
PeuCap = ((Capacity/10)^k)*10; % See equation 2.12

% Starting voltage set outside loop
E=open_circuit_voltage_LA(0,NoCells);
I = (E - (E*E - (4*Rin*P))^0.5)/(2*Rin);    %Equation 2.20
V(1) = E - I*Rin;                            %Equation 2.1
```

---

[6] As with all quadratics, there are two solutions. The other corresponds to a 'lunatic' way of operating the battery at a huge current, so large that the internal resistance causes the voltage to drop to a low value, so that the power is achieved with a low voltage and very high current. This is immensely inefficient.

Batteries

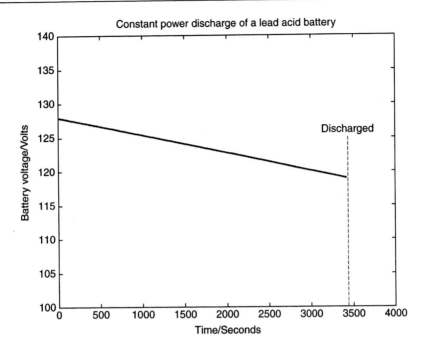

Figure 2.17 Graph of voltage against time for a constant power discharge of a lead acid battery at 5000 W. The nominal ratings of the battery are 120 V, 50 Ah

```
for n=2:1001
   E=open_circuit_voltage_LA(DoD(n-1),NoCells);   %Equ 2.10
   I = (E - (E*E - (4*Rin*P))^0.5)/(2*Rin);       %Eq 2.20
   CR(n) = CR(n-1) + ((deltaT * I^k)/3600);       %Eq 2.17
   DoD(n) = CR(n)/PeuCap;                         %Eq 2.19
   if DoD(n)>1
      DoD(n)=1;
   end
   % We will say that the battery is "dead" if the
   % depth of discharge exceeds 99%
   V(n)=open_circuit_voltage_LA(DoD(n),NoCells) - I*Rin;  %Equ 2.1
   if DoD(n)>0.99
      V(n)=0;
   end
end
plot(T,V,'b.');
YLABEL('Battery voltage/Volts');
XLABEL('Time/Seconds');
TITLE('Constant power discharge of a lead acid battery');
axis([0 4000 100 140]);
```

When we come to simulate the battery being used in a vehicle, the issue of regenerative braking will arise. Here a certain power is dissipated *into* the battery. If we look again at Figure 2.1, and consider the situation that the current I is flowing *into* the battery, then the equation becomes:

$$V = E + IR \tag{2.21}$$

If we combine equation (2.21) with the normal equation for power we obtain:

$$P = V \times I = (E + IR) \times I = EI + RI^2$$

The 'sensible', normal efficient operation, solution to this quadratic equation is:

$$I = \frac{-E + \sqrt{E^2 + 4RP}}{2R} \tag{2.22}$$

The value of R, the internal resistance of the cell, will normally be different when charging as opposed to discharging. To use a value twice the size of the discharge value is a good first approximation.

When running a simulation, we must remember that the power P is positive, and that equation (2.22) gives the current *into* the battery. So when incorporating regenerative braking into battery simulation, care must be taken to use the right equation for the current, and that equation (2.17) must be modified so that the charge *removed* from the battery is *reduced*. Also, it is important to remove the Peukert Correction, as when charging a battery large currents do not have proportionately more effect than small ones. Equation (2.17) thus becomes:

$$CR_{n+1} = CR_n - \frac{\delta t \times I}{3600} \text{Ah} \tag{2.23}$$

We shall meet this equation again in Chapter 7, Section 7.4.2, where we simulate the range and performance of electric vehicles with and without regenerative braking.

2.11.5 Calculating the Peukert Coefficient

These equations and simulations are very important, and will be used again when we model the performance of electric vehicles in Chapter 7. There the powers and currents will not be constant, as they were above, but exactly the same equations are used.

However, all this begs the question 'How do we find out what the Peukert Coefficient is?' It is very rarely given on a battery specification sheet, but fortunately there is nearly always sufficient information to calculate the value. All that is required is the battery capacity at two different discharge times. For example, the nominally 42 Amphours (10 hour rating) battery of Figure 2.2 also has a capacity of 33.6 Amphours at the 1 hour rate.

The method of finding the Peukert Coefficient from two amphour ratings is as follows. The two different ratings give two different rated currents:

$$I_1 = \frac{C_1}{T_1} \quad \text{and} \quad I_2 = \frac{C_2}{T_2} \tag{2.24}$$

Batteries

We then have two equations for the Peukert Capacity, as in equation (2.12):

$$C_p = I_1^k \times T_1 \quad \text{and} \quad C_p = I_2^k \times T_2 \tag{2.25}$$

However, since the Peukert Coefficient is Constant, the right hand sides of both parts of equation (2.25) are equal, and thus:

$$I_1^k T_1 = I_2^k T_2$$

$$\left(\frac{I_1}{I_2}\right)^k = \frac{T_2}{T_1}$$

Taking logs, and rearranging this gives:

$$k = \frac{(\log T_2 - \log T_1)}{(\log I_1 - \log I_2)} \tag{2.26}$$

This equation allows us to calculate the Peukert Coefficient k for a battery, provided we have two values for the capacity at two different discharge times T. Taking the example of our 42 Ah nominal battery, equation (2.24) becomes:

$$I_1 = \frac{C_1}{T_1} = \frac{42}{10} = 4.2 \text{ A} \quad \text{and} \quad I_2 = \frac{C_2}{T_2} = \frac{33.6}{1} = 33.6 \text{ A}$$

Putting these values into equation (2.26) gives:

$$k = \frac{\log 1 - \log 10}{\log 4.2 - \log 33.6} = 1.107$$

Such calculations can be done with any battery, provided some quantitative indication is given as to how the capacity changes with rate of discharge. If a large number of measurements of capacity at different discharge times are available, then it is best to plot a graph of $\log(T)$ against $\log(I)$. Clearly, from equation (2.26), the gradient of the best-fit line of this graph is the Peukert Coefficient.

As a general rule, the lower the Peukert Coefficient, the better the battery. All battery types behave in a similar way, and are quite well modelled using this method. The Peukert Coefficient tends to be rather higher for the lead acid batteries than for other types.

2.11.6 Approximate battery sizing

The modelling techniques described above, when used with the models for vehicles described in Chapter 7, should be used to give an indication of the performance that will be obtained from a vehicle with a certain type of battery. However, it is possible, and sometimes useful, to get a very approximate guide to battery range and/or size using the approach outlined below.

A designer may either be creating a new vehicle or alternatively may be adapting an existing vehicle to an electric car. The energy consumption of an existing vehicle will probably be known, in which case the energy used per kilometre can be multiplied by the range and divided by the specific energy of the battery to give an approximate battery mass.

If the vehicle is a new design the energy requirements may be obtained by comparing it with a vehicle of similar design. Should the similar vehicle have an IC engine, the energy consumption can be derived from the fuel consumption and the engine/gear box efficiency.

This method is fairly crude, but none the less may give a reasonable answer which can be analysed later.

For example the vehicle may be compared with a diesel engine car with a fuel consumption of $18\,\text{km.l}^{-1}$ (50 mpg). The specific energy of diesel fuel is approximately $40\,\text{kWh.kg}^{-1}$ and the conversion efficiency of the engine and transmission is approximately 10%, resulting in 4 kWh of energy per litre of fuel stored delivered at the wheels.

In order to travel 180 km the vehicle will consume 10 litres of fuel, which weights approximately 11 kg allowing for fuel density. This fuel has an energy value of 440 kWh, and the energy delivered to the wheels will be 44 kWh (44 000 Wh) allowing for the 10% efficiency. This can be divided by the electric motor and transmission efficiency, typically about 0.7 (70%), to give the energy needed from the battery, i.e. 62.8 kWh or 62 800 Wh.

Hence if a lead acid battery is used (specific energy $35\,\text{Wh.kg}^{-1}$) the battery mass will be 1257 kg; if a NiMH battery (specific energy $60\,\text{Wh.kg}^{-1}$) is used the battery mass will be 733 kg; if a sodium nickel chloride battery is used of a specific energy of $86\,\text{Wh.kg}^{-1}$ then the battery mass will be 511 kg; and if a zinc air battery of $230\,\text{Wh.kg}^{-1}$ is used a battery mass of 191 kg is needed.

Care must be used when using specific energy figures particularly, for example, when a battery such as a lead acid battery is being discharged rapidly, when the specific energy actually obtained will be considerably lower than the nominal $35\,\text{Wh.kg}^{-1}$. However this technique is useful, and in the case quoted above gives a fairly good indication of which batteries would be ideal, which would suffice and which would be ridiculously heavy. The technique would also give a 'ball park figure' of battery mass for more advanced analysis, using the modelling techniques introduced above, and much further developed in Chapter 7.

2.12 In Conclusion

There have been massive improvements in batteries in recent years, and several new developments are showing considerable promise. Nevertheless the specific energies of batteries, with the possible exception of zinc-air, are still extremely low.

A fairly standard four door motor car travelling at 100 km/h may typically use an average power of 30 kW. The mass of different types of battery for different distances travelled are shown in Table 2.11, assuming an electric motor/drive efficiency of 0.7 (70%).

Although these are approximate figures they are comparative, and they do give a fairly clear picture as to the state of battery development for electric vehicles.

Lead acid batteries are only really suited for short-range vehicles. They remain the cheapest form of battery per unit of energy stored and it is likely that they will continue to be widely used for these purposes. Very many useful electric vehicles can be made which do not need a long range.

Table 2.11 Examples of distance travelled/battery weight for a typical car

Battery type	Specific energy Wh.kg^{-1}	Battery mass kg, 75 km range	Battery mass kg, 150 km range	Battery mass kg, 225 km range	Battery mass kg, 300 km range
Lead acid	30	750	1500	2250	3000
NiMH	65	346	692	1038	1385
Li ion	90	250	500	750	1000
NaNiCl	100	225	450	675	900
Zn-Air	230	98	196	293	391

Some of the newer batteries, such as nickel metal hydride, lithium ion and particularly sodium nickel chloride, have sufficient energy density to be used for medium range vehicles. More importantly, batteries such as nickel metal hydride can be charged very rapidly which makes them ideal for use in hybrid cars, with range extenders, or for a vehicle such as a bus or tram which can be recharged during frequent stops. These batteries are currently expensive, particularly the lithium ion battery. Provided that predicted price decreases are accurate, it is likely that these batteries will one day become widely used.

There are no batteries that currently show signs of enabling pure electric vehicles to compete in both versatility and long-range use with IC engine vehicles. To do this a totally different technology is needed, which leads us into the consideration of fuel cells in Chapters 4 and 5. However, before we consider such radical technology, we take a look at other ways of storing electrical energy, apart from batteries.

References

Chou Y.-F., Peng K.-K., Huang M.-F., Pun H.-Y., Lau C.-S., Yang M.-H. and Shuy G.W. (2001) A battery management system of electric scooter using Li-ion battery pack. *Proceedings of the 18th International Electric Vehicle Symposium*, CD-ROM.

Johnson V.H., Zolot M.D. and Pesaran A.A. (2001) Development and validation of a temperature-dependent resistance/capacitance model for ADVISOR. *Proceedings of the 18th International Electric Vehicle Symposium*, CD-ROM.

Schmitz P. and Busch R. (2001) System integration and validation of a Li-ion battery in an advanced electric vehicle. *Proceedings of the 18th International Electric Vehicle Symposium*, CD-ROM.

Shnayerson M. (1996) *The Car That Could*, Random House, New York.

Sweigert G.M., Eley K. and Childers C. (2001) Standardisation of charging systems for battery electric vehicles. *Proceedings of the 18th International Electric Vehicle Symposium*, CD-ROM.

Vincent C.A. and Scrosati B. (1997) *Modern Batteries*, Arnold, London.

3

Alternative and Novel Energy Sources and Stores

3.1 Introduction

In addition to conventional electrical power sources for electric vehicles such as batteries and fuel cells, there is a range of alternative options including solar photovoltaics, wind-driven generators, flywheels and supercapacitors. There are also older systems which may be important in the development of electric vehicles, particularly electric supply rails either with mechanical pick-ups or modern ones with an inductive supply.

In this chapter we are considering *stores* of electrical energy, energy *conversion* devices, and energy *transfer* systems. The common feature is that they all seek to supply electricity to autonomous vehicles, in ways other than the batteries described in the last chapter, and the fuel cells to be described in Chapter 4.

3.2 Solar Photovoltaics

Photovoltaic cells are devices that convert sunlight or solar energy into direct current electricity. They are usually found as flat panels, and such panels are now a fairly common sight, on buildings and powering roadside equipment, to say nothing of being on calculators and similar electronic equipment. They can also come as thin films, which can be curved around a car body.

Solar radiation strikes the upper atmosphere with a value of $1300\,\mathrm{Wm}^{-2}$ but some of the radiation is lost in the atmosphere and by the time it reaches the Earth's surface it is less than $1000\,\mathrm{Wm}^{-2}$, normally called a 'standard sun'. Even in hot sunny climates solar radiation is normally less than this. Typical solar radiation on a flat plate constantly turned towards the sun will average around $750\,\mathrm{Wm}^{-2}$ on a clear day in the tropics and around $500\,\mathrm{Wm}^{-2}$ in more hazy climates such as the Philippines. For a flat plate such as a solar panel placed on a car roof, the sun will strike the plate at differing angles as the sun moves around the sky, which halves the amount of energy falling on the plate. That is, the average energy falling on a flat surface such as a car roof would be $375\,\mathrm{Wm}^{-2}$

Electric Vehicle Technology Explained   James Larminie and John Lowry
© 2003 John Wiley & Sons, Ltd   ISBN: 0-470-85163-5

for a clear day in the tropics and $250\,\text{Wm}^{-2}$ in the Philippines. The exact average will depend on the latitude, being larger on the equator and less at higher latitudes.

Solar radiation is split into direct radiation which comes from the direction of the sun which is normally prominent on cloudless days, and indirect radiation which is solar radiation broken up by cloud and dust, comes from all directions and is prominent on cloudy days. Photovoltaic cells convert both types of radiation into electricity with an efficiency of conversion of around 14%. So the power which could be obtained from a photovoltaic panel will be less than $100\,\text{Wm}^{-2}$ when tracking the sun, and around half of this for a fixed panel on a horizontal car roof.

There are two methods of using solar panels, either onboard or offboard the vehicle. Clearly even if the whole of a car plan area were covered with cells only a very limited amount of power would be obtained. For example, a car of plan area $5\,\text{m}^2$ would produce a maximum of around 375 W at the panel output, and an average of around 188 W, giving 1.88 kWh of energy over a 10 hour day, equivalent to the energy stored in around 50 kg of lead acid batteries. This energy could be stored in a battery and used to power the vehicle for short commuter and shopping trips; but basically this amount of energy is insubstantial and would normally only give an impracticably limited range. BP uses a solar powered vehicle for use around its factory in Madrid and solar powered golf buggies are currently being developed. An annual race of vehicles powered by solar photovoltaics is held in Australia. Although vehicles have achieved speeds of 85 kph (50 mph) across Australia they are not really appropriate for everyday use. They have large surface areas, are normally made from bicycle components, have very limited interior space and require very high levels of solar radiation, hence the race being held in Australia.

Solar panels mounted offboard could give as much power as needed. The electricity could either charge the vehicle battery from a suitable charging point or could be supplied to the vehicle via supply rails. The idea of a solar roof could be wasteful, in the sense that it is expensive and when the car is not being used the power will go to waste, unless of course it is used for some other purpose such as charging domestic batteries at a remote residence. The surplus power could be sold back to the grid in some cases.

Apart from the disadvantage of low power per square metre, solar panels are not cheap, costing around £4000 per peak kW, when bought in bulk. Bearing in mind that a peak kW is rarely achieved even in very sunny places, the actual cost per kW achieved is considerably more than £4000. Despite this, the idea of solar photovoltaics fitted to vehicles should not be written off entirely. The efficiency will improve and may one day in the future be as high as 50%. The cost of photovoltaics has already fallen dramatically and the long-term cost of solar photovoltaic panels is predicted to fall still further. Apart from supplying power to drive the vehicle, solar photovoltaics may be used for other useful purposes, such as compensating for natural battery self-discharge, and also for cooling or heating the car whilst at rest. A small fan powered by a photovoltaic roof panel could be used to draw air through a vehicle and keep it cool when parked in the sun. Examples of designs of vehicles featuring integral solar panels and analyses of the possibilities can be found in Kumagi and Tatemoto (1989) and Fujinnaka (1989).

Using solar energy to energise vehicles from the mains grid is discussed further in Chapter 10, where it is shown that solar power should eventually contribute as a clean sustainable power source for the future.

3.3 Wind Power

Wind-driven electric generators can also be used to charge the batteries of stationary electric vehicles. Stationary wind generators, such as smaller versions of the one illustrated in Figure 10.6, are common methods of supplying power in areas without mains electricity.

A stationary wind generator could be used in the same way as a stationary photovoltaic array. Alternatively it would be possible to mount a small generator on the roof of an electric vehicle, for charging when the vehicle was stationary. There would be no point in using it when the vehicle was in motion, as the power gained from the wind generator would be considerably less than the power lost by dragging the wind generator through the wind, the efficiency being less than 100%. Ideally, for aerodynamic reasons the wind generator would fold away when the vehicle was travelling. The concept of an onboard wind generator is illustrated in Figure 3.1. In windy places a wind generator 1.2 m in diameter could produce up to 500 W continuously whilst the wind speed averaged $10\,\text{ms}^{-1}$. Whether such an idea is practical for general use is debatable.

The power from the wind has a similar energy per square metre as solar radiation. The actual power P in W is given by the formula:

$$P = 0.5 \rho A v^3 \qquad (3.1)$$

where ρ is the air density (kgm^{-3}), v is the wind speed (ms^{-1}) and A is the area (m^2) through which the wind passes. Hence with a $10\,\text{ms}^{-1}$ wind the power is $625\,\text{Wm}^{-2}$,

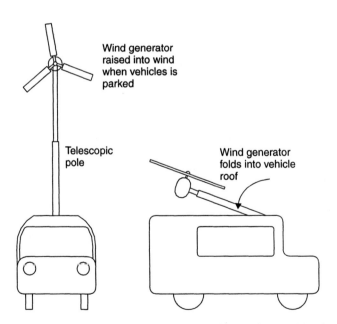

Figure 3.1 Concept of onboard wind generator for charging, which would only be practical in very limited circumstances

assuming an air density of $1.25\,\text{kgm}^{-3}$. The amount of electrical power realised is typically around 30% of this. (It is governed by the theoretical Betz efficiency and the relative efficiency of the wind/electric generator in question.)

Solar and wind energy can be used in conjunction. The potential of wind energy being used for transport when supplied to the grid is discussed in greater detail in Chapter 10.

3.4 Flywheels

Flywheels are devices that are used for storing energy. A plane disc spinning about its axis would be an example of a simple flywheel. The kinetic energy of the spinning disc is released when the flywheel slows down. The energy can be captured by connecting an electrical generator directly to the disc as shown in Figure 3.2, power electronics being required to match the generator output to a form where it can drive the vehicle motors. The flywheel can be re-accelerated, acting as a regenerative brake. Alternatively the flywheel can be connected to the vehicle wheels via a gearbox and a clutch. A photograph of a flywheel and purely mechanical transmission used on the Parry People Mover is shown in Figure 3.3. This transmission matches the rotational speed of the flywheel to the wheels, the flywheel giving out energy as it decelerates.

Whether mechanical or electrical, the system can also be used to recover kinetic energy when braking. The flywheel can be accelerated, turning the kinetic energy of the vehicle into stored kinetic energy in the flywheel, and acting as a highly efficient regenerative brake.

The total amount of energy stored is given by the formula:

$$E = 0.5I\omega^2 \qquad (3.2)$$

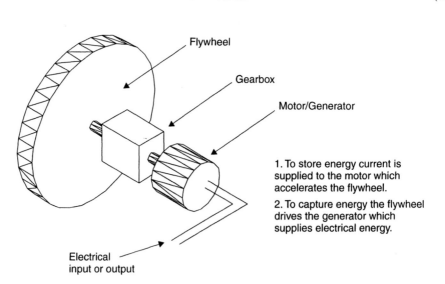

Figure 3.2 Flywheel/generator arrangement

Figure 3.3 Parry People Mover chassis. The enclosed flywheel can be clearly seen in the middle of the vehicle (Photograph kindly supplied by Parry People Movers Ltd.)

where E is the energy in joules, I is the moment of inertia and ω is the rotational speed in radians per second. When a flywheel reduces from ω_1 to ω_2 rad s^{-1} the energy released will be given by the formula:

$$\Delta E = 0.5 I (\omega_1^2 - \omega_2^2) \tag{3.3}$$

If you could make a flywheel strong enough almost infinite energy could be stored, bearing in mind that the mass and hence the moment of inertia get larger as the flywheel peripheral speed approaches the speed of light. Unfortunately as the flywheel rotational speed increases so do the stresses in the material. As a result the flywheel's energy storage capacity is limited by the tensile strength of the material it is made from.

The main advantage of flywheels is that they have a high specific power and it is relatively easy to get energy to and from the flywheel. They are also fairly simple, reliable mechanical devices. The specific energy from flywheels is limited and unlikely to approach that of even lead acid batteries. Attempts have been made to boost specific energy by using ultra-strong materials, running the flywheel in inert gas or a vacuum to reduce air friction losses, and using magnetic bearings.

Apart from the low specific energy there are major worries about safety due to the risk of explosion. In the event of the flywheel rupturing, during a crash energy is released almost instantly and the flywheel effectively acts like a bomb. Also, if a fast moving flywheel becomes detached from its mountings it could cause real havoc. Another aspect of flywheels that needs to be considered is the gyroscopic effect of a disc rotating at high speeds. Firstly, without outside interference they tend to stay in one position and do not readily move on an axis other than the axis of spin. When a torque or movement is introduced around one axis, the flywheel tends to move or precess around another axis. Again the behaviour in an accident situation needs to be studied carefully, as does the effect on the vehicle's dynamics. However, it should be noted that in many cases these effects could be benign, and they could have a smoothing effect on vehicle ride.

Several attempts have been made to produce flywheel buses and trams, the Parry People Mover of Figure 1.17 being, we believe, the only one available commercially. The chassis from this vehicle, showing the flywheel device, is shown in Figure 3.3.

Despite the lack of success of the flywheel for vehicle energy storage and a certain amount of bad press, it would be wrong to write off the flywheel completely. Virtually all IC engines have small flywheels and these have not proved particularly problematic. The simplicity of a small flywheel to be used in an electric vehicle for use as a regenerative braking system should not be overlooked. Provided the flywheel is used well below its rupture point and is kept relatively small and well guarded, it may come to have a useful role in the future of electric vehicles, particularly in hybrids.

3.5 Super Capacitors

Capacitors are devices in which two conducting plates are separated by an insulator. An example is shown in Figure 3.4. A DC voltage is connected across the capacitor, one plate

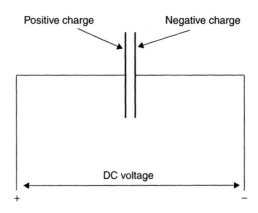

Figure 3.4 Principle of the capacitor

being positive the other negative. The opposite charges on the plates attract and hence store energy. The charge Q stored in a capacitor of capacitance C Farads at a voltage of V Volts is given by the equation:

$$Q = C \times V \qquad (3.4)$$

As with flywheels, capacitors can provide large energy storage, although they are more normally used in small sizes as components in electronic circuits. The large energy-storing capacitors with large plate areas have come to be called super capacitors. The energy stored in a capacitor is given by the equation:

$$E = \tfrac{1}{2}CV^2 \qquad (3.5)$$

where E is the energy stored in Joules. The capacitance C of a capacitor in Farads will be given by the equation:

$$C = \varepsilon \frac{A}{d} \qquad (3.6)$$

where ε is the is the permittivity of the material between the plates, A is the plate area and d is the separation of the plates. The key to modern super capacitors is that the separation of the plates is so small. The capacitance arises from the formation on the electrode surface of a layer of electrolytic ions (the double layer). They have high surface areas, e.g. $1\,000\,000\,m^2 kg^{-1}$, and a 4000 F capacitor can be fitted into a container the size of a beer can.

However, the problem with this technology is that the voltage across the capacitor can only be very low, between 1 to 3 V. The problem with this is clear from equation (3.5), it severely limits the energy that can be stored. In order to store charge at a reasonable voltage many capacitors have to be connected in series. This not only adds cost, it brings other problems too.

If two capacitors C_1 and C_2 are connected in series then it is well known[1] that the combined capacitance C is given by the formula:

$$\frac{1}{C} = \frac{1}{C_1} + \frac{1}{C_2} \qquad (3.7)$$

So, for example, two 3 F capacitors in series will have a combined capacitance of 1.5 F. Putting capacitors in series *reduces* the capacitance. Now, the energy stored increases as the voltage *squared*, so it does result in more energy stored, but not as much as might be hoped from a simple consideration of equation (3.5).

Another major problem with putting capacitors in series is that of charge equalisation. In a string of capacitors in series the charge on each one should be the same, as the same current flows through the series circuit. However, the problem is that there will be a certain amount of self-discharge in each one, due to the fact that the insulation between the plates of the capacitors will not be perfect. Obviously, this self-discharge will not be equal in

---

[1] Along with all the equations in this section, a fully explanation or proof can be found in any basic electrical circuits or physics textbook.

all the capacitors; life is not like that! The problem then is that there may be a relative charge build-up on some of the capacitors, and this will result in a higher voltage on those capacitors. It is certain that unless something is done about this, the voltage on some of the capacitors will exceed the maximum of 3 V, irrevocably damaging the capacitor.

This problem of voltage difference will also be exacerbated by the fact that the capacitance of the capacitors will vary slightly, and this will affect the voltage. From equation (3.4) we can see that capacitors with the same charge and different capacitances will have different voltages.

The only solution to this, and it is essential in systems of more than about six capacitors in series, is to have *charge equalisation circuits*. These are circuits connected to each pair of capacitors that continually monitor the voltage across adjacent capacitors, and move charge from one to the other in order to make sure that the voltage across the capacitors is the same.

These charge equalisation circuits add to the cost and size of a capacitor energy storage system. They also consume some energy, though designs are available that are very efficient, and which have a current consumption of only 1 mA or so. A good review and more detailed explanation of equalisation circuits in capacitor storage systems can be found in Härri and Egger (2001).

A super capacitor energy storage system is shown in Figure 3.5. In this picture, which is of the capacitors used in the bus that was shown in Figure 1.18, we can see capacitors

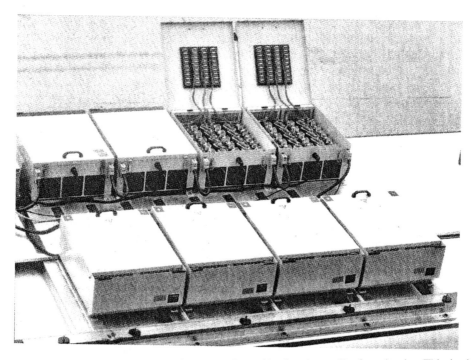

Figure 3.5 A bank of super capacitors, together with charge equalisation circuits. This is the system from the bus of Figure 1.18 (photograph kindly supplied by MAN Nutzfahrzeuge AG.)

Alternative and Novel Energy Sources and Stores

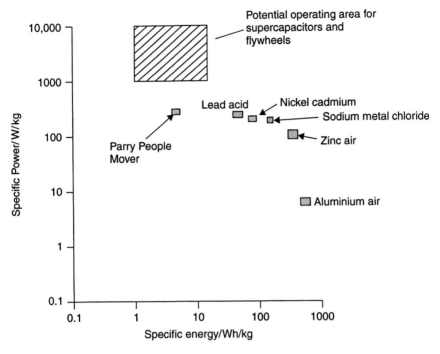

Figure 3.6 Ragone plot of batteries, supercapacitors and flywheels

connected in series, and also the charge equalisation circuits mentioned above. A Ragone plot comparing supercapacitors with batteries is shown in Figure 3.6.

In many ways the characteristics of supercapacitors are like those of flywheels. They have relatively high specific power and relatively low specific energy. They can be used as the energy storage for regenerative braking. Although they could be used alone on a vehicle, they would be better used in a hybrid as devices for giving out and receiving energy rapidly during braking and accelerating afterwards, e.g. at traffic lights. Supercapacitors are inherently safer than flywheels as they avoid the problems of mechanical breakdown and gyroscopic effects. Power electronics are needed to step voltages up and down as required. Several interesting vehicles have been built with super capacitors providing significant energy storage, and descriptions of these can be found in the literature. Furubayashi *et al.* (2001) describe a system where capacitors are used with a diesel IC engine. Lott and Späth (2001) describe a capacitor/zinc-air battery hybrid, and Büchi *et al.* (2002) describe a system where capacitors are used with a fuel cell.

3.6 Supply Rails

Electrical supply rails date from the 19th century and are an old and well tested method of supplying electric vehicles. Originally they were used on railed vehicles and later on trolley buses.

The advantage of supply rails is that an electrical vehicle can be used without the need for an onboard battery. This enables clean, non-polluting vehicles to be used that can have an almost infinite range. This is ideal for underground trains for example. Originally the London underground used steam trains! The disadvantage of supply rails is that the vehicle has to follow a pre-determined route. Trolley buses are normally fitted with small batteries that allow them to be driven for a short range away from the supply. The latter is a valuable idea. For example, electric vehicles could run on roads or tracks using supply rails on specific routes and then run off the tracks using batteries or fuel cells when starting and completing a journey.

Mechanical supply rails have several disadvantages. Firstly they rely on two surfaces rubbing together and as a result they wear and need appropriate maintenance. Secondly there is a tendency for arcing, which causes further wear and is off-putting to vehicle passengers.

A more modern approach is to use inductive pick-up rails. An example of a modern commercial system is the IPT (Inductive Power Transfer) discussed below.

The system was conceived in the University of Auckland, New Zealand, and the development of an IPT product is being carried out commercially by a company called Wampfler, who own the patent.

IPT is a contactless power supply system that would allow electrical energy to be safely supplied to vehicles without any mechanical contact. IPT works by the same principle as a transformer. The primary circuit lies on the track whilst the pick-up is the secondary. In a regular transformer the airgap between primary and secondary is very small and the frequency is low (50/60 Hz). With IPT the airgap is large but the operating frequency is raised to 15 000 Hz to compensate. With the large airgap the system becomes insensitive to positional tolerances of the pick-up on the track. Multiple loads may also be operated at the same time. The track power supply generates the high-frequency alternating current in the track cable. The special shape of the pick-up is most effective at capturing the magnetic field generated by the track conductors. The captured AC magnetic field produces electrical energy in the pick-up coil and the pick-up regulator converts the high frequency AC power to DC while regulating the power to the load. If required the DC can be converted back to AC at a chosen frequency using an inverter.

IPT may be used to continuously supply electrical energy along a predetermined track to people movers such as monorails, duo-rails, or elevators, as well as theme park rides. The main features of the IPT system are:

- Efficiency: the track power supply and vehicle pick-up work with an efficiency of up to 96%. Both track and pick-up systems are resonant so that losses and harmonics are minimised.
- Power: hundreds of kWs may be transferred. Power ranges of 30–1000 kW are possible.
- Large Airgap: power may be transferred across airgaps of 100 mm and more.
- Multiple Independent Loads: using intelligent control, any number of vehicles may be operated independently and simultaneously on a system.
- Long Tracks: IPT works with track lengths of up to several kilometres in length, which may be repeated for even longer systems.

- Maintenance: no brush wear or moving parts ensure that the IPT system is virtually maintenance-free.
- Data Transfer: signal and data transfer is possible with IPT with minimal additional hardware. An integrated power and data system is currently being developed.
- Speed: with conductor bar, festoon or cable reel systems, speed is a limiting factor. With IPT speed of operation is unlimited.
- Safety: all components are fully enclosed and insulated. Hence the system is fully touch-proof.
- Sensitive Environments: the fact that no carbon dust, other wear or sparks are generated make IPT suitable for sensitive or hazardous environments.

The company, Wampfler, have built a test track at its headquarters in Weil am Rhein in Germany. It is claimed, to be the largest IPT system constructed to date, having a total capacity of 150 kW and a track length of nearly 400 m. Power is transferred to a total of 6 pick-ups on a test vehicle, each having a power capability of 25 kW and an airgap of 120 mm. Taking into account the track cable and the track supports, this allows a positional tolerance of movement of the pick-up of 50 mm in all directions. Since the IPT test vehicle requires a peak power no greater than 10 kW, the excess power is returned via conductor bar for regeneration into the mains. The test track will be used as a basis for the development of the product range and for the continued analysis of cost optimisation.

The IPT system could be used for buses and cars. It can operate with either an enclosed style pick-up as illustrated above, or with a flat roadway style pick-up. The track conductors may be buried in the roadway or in the charging station platform and a flat pick-up is used as an energy collector. The flat construction allows large lateral tolerances. The energy transfer is totally contactless and intervention free.

In the future many other application areas may be covered using IPT, including trams and underground trains.

This IPT system is also potentially applicable to hybrid vehicles. The use of electric vehicles, which take power from supply rails within cities and on motorways, could itself cause a revolution in electric transport.

A system was proposed by the author in which electric cars can be driven on normal roads using power from their batteries and also use special tracks in cities and on motorway routes. Whilst on the track the vehicles would be powered from a supply rail or inductive pick-up, the batteries being recharged at the same time. The cars on the track would be under total automatic control. The tracks would be guarded from pedestrians and stray animals.

Such a system has considerable advantages. Vehicles could use electric batteries to take them to and from the track and for short journeys. They would use the special tracks, using power from the pick up rails and charging their batteries at the same time. Because vehicles travelling on the track will travel at a constant speed they will use less energy than conventional city traffic, which keeps stopping and starting. Vehicles in cities such as London and Tokyo have such low average speeds for much of the day, e.g. 12 kph that even with relatively slow track speeds, e.g. 48 kph, the speed of movement would be increased four-fold.

References

Büchi F., Tsukada A., Rodutz P., Garcia O., Ruge M., Kötz R., Bärtschi M. and Dietrich P. (2002) Fuel cell supercap hybrid electric power train. *The Fuel Cell World 2002*, Proceedings, European Fuel Cell Forum Conference, Lucerne, pp. 218–231.

Fujinnaka M. (1989) Future vehicles will run with solar energy. Proceedings of an SAE conference published as *Recent Advances in Electric Vehicle Technology*, Society of Automotive Engineers, pp. 31–39.

Furubayashi M., Ushio Y., Okumura E., Takeda T., Andou D. and Shibya H. (2001) Application of high power super capacitors to an idling stop system for city buses. *Proceedings of the 18th International Electric, Fuel Cell and Hybrid Vehicles Symposium*, CD-ROM.

Härri V.V. and Egger S. (2001) Supercapacitor circuitry concept SAM for public transport and other applications. *Proceedings of the 18th International Electric, Fuel Cell and Hybrid Vehicles Symposium*, CD-ROM.

Kumagi N. and Tatemoto M. (1989) Application of solar cells to the automobile. Proceedings of an SAE conference published as *Recent Advances in Electric Vehicle Technology*, Society of Automotive Engineers, pp. 117–121.

Lott J. and Späth H. (2001) Double layer capacitors as additional power sources in electric vehicles. *Proceedings of the 18th International Electric, Fuel Cell and Hybrid Vehicles Symposium*, CD-ROM.

4

Fuel Cells

4.1 Fuel Cells, a Real Option?

Fuel cells are hardly a new idea. They were invented in about 1840, but they are yet to really make their mark as a power source for electric vehicles. However, this might be set to change over the next 20 or 30 years. Certainly most of the major motor companies are spending very large sums of money developing fuel cell powered vehicles.

The basic principle of the fuel cell is that it uses hydrogen fuel to produce electricity in a battery-like device to be explained in the next section. The basic chemical reaction is:

$$2H_2 + O_2 \longrightarrow 2H_2O \tag{4.1}$$

The product is thus water, and energy. Because the types of fuel cell likely to be used in vehicles work at quite modest temperatures ($\sim 85°C$) there is no nitrous oxide produced by reactions between the components of the air used in the cell. A fuel cell vehicle could thus be described as zero-emission. Furthermore, because they run off a fairly normal chemical fuel (hydrogen), very reasonable energies can be stored, and the range of fuel cell vehicles is potentially quite satisfactory. They thus offer the only real prospect of a silent zero-emission vehicle with a range and performance broadly comparable with IC engined vehicles. It is not surprising then that there have, for many years, been those who have seen fuel cells as a technology that shows great promise, and could even make serious inroads into the domination of the internal combustion engine. Such ideas regularly surface in the science and technology community, and Figure 4.1, showing a recent cover of the prestigious *Scientific American* magazine, is but one example.

Many demonstration fuel cell powered cars of very respectable performance have been made, and examples are shown in Figures 1.14 and 1.15. However, there are many problems and challenges for fuel cells to overcome before they become a commercial reality as a vehicle power source. The main problems centre around the following issues.

1. Cost: fuel cells are currently far more expensive than IC engines, and even hybrid IC/electric systems. The reasons for this are explained in Section 4.4, where we consider how a fuel cell system is made, and in Section 4.7, where we show the extent of the equipment that needs adding to a fuel cell to make a working system.

Electric Vehicle Technology Explained   James Larminie and John Lowry
© 2003 John Wiley & Sons, Ltd   ISBN: 0-470-85163-5

Figure 4.1 The front cover of the October 2002 issue of the magazine Scientific American. Articles within outline the possibilities presented by fuel cell powered electric vehicles. (Reproduced by kind permission of Scientific American.)

2. Rival technology: hydrogen is a fuel, and it can be used with exactly the same overall chemical reaction as equation (1.1) in an IC engine. Indeed, cars have been produced with fairly conventional engines running off hydrogen, notably by BMW in Germany. The emissions from these vehicles are free from carbon monoxide, carbon dioxide, hydrocarbons, and virtually all the unpleasant pollution associated with cars; the only pollutant is a small amount of nitrous oxide. Considering the reduced cost and complexity, is this a better solution? To answer this question we need to look at the efficiency of a fuel cell, and see how it compares with an IC engine. This basic thermodynamics are covered in Section 4.3.
3. Water management: it is not at all self-evident why water management should be such an important and difficult issue with automotive fuel cells, so Section 4.5 is devoted to explaining this important and difficult problem.
4. Cooling: the thermal management of fuel cells is actually rather more difficult than for IC engines. The reasons for this, and the solutions, are discussed in Section 4.6.
5. Hydrogen supply: hydrogen is the preferred fuel for fuel cells, but hydrogen is very difficult to store and transport. There is also the vital question of 'where does the hydrogen come from?' These issues are so difficult and important, with so many rival solutions, that we have dedicated a whole chapter to them, Chapter 5.

Fuel Cells

Figure 4.2 A fuel cell powered bus in use in Germany. Vehicles like this, used all day in cities, and refueling at one place, are particularly suited to being fuel cell powered (Reproduced by kind permission of MAN Nutzfahrzeuge AG.)

However, there is great hope that these problems can be overcome, and fuel cells can be the basis of less environmentally damaging transport. Many of the problems are more easily solved, and the benefits are more keenly felt, with vehicles such as buses that run all day in large cities. Such a vehicle is shown in Figure 4.2, and they have been used in several major cities in Canada, the USA and Europe. Many thousands of people will have taken journeys in a fuel cell powered vehicle, though many of them will not have noticed it. So before we consider the major problems with fuel cells, we will explain how these interesting devices work.

4.2 Hydrogen Fuel Cells: Basic Principles

4.2.1 Electrode reactions

We have seen that the basic principle of the fuel cell is the release of energy following a chemical reaction between hydrogen and oxygen. The key difference between this and simply burning the gas is that the energy is released as an electric current, rather that heat. How is this electric current produced?

To understand this we need to consider the separate reactions taking place at each electrode. These important details vary for different types of fuel cell, but if we start with a cell based on an acid electrolyte, we shall consider the simplest and the most common type.

At the anode of an acid electrolyte fuel cell the hydrogen gas ionises, releasing electrons and creating H^+ ions (or protons).

$$2H_2 \longrightarrow 4H^+ + 4e^- \qquad (4.2)$$

This reaction releases energy. At the cathode, oxygen reacts with electrons taken from the electrode, and H^+ ions from the electrolyte, to form water.

$$O_2 + 4e^- + 4H^+ \longrightarrow 2H_2O \qquad (4.3)$$

Clearly, for both these reactions to proceed continuously, electrons produced at the anode must pass through an electrical circuit to the cathode. Also, H^+ ions must pass through the electrolyte. An acid is a fluid with free H^+ ions, and so serves this purpose very well. Certain polymers can also be made to contain mobile H^+ ions. These materials are called 'proton exchange membranes', as an H^+ ion is also a proton, and their construction is explained below in Section 4.5.

Comparing equations (4.2) and (4.3) we can see that two hydrogen molecules will be needed for each oxygen molecule if the system is to be kept in balance. This is shown in Figure 4.3. It should be noted that the electrolyte must allow only H^+ ions to pass through it, and not electrons. Otherwise the electrons would go through the electrolyte, not round the external circuit, and all would be lost.

4.2.2 Different electrolytes

The reactions given above may seem simple enough, but they do not proceed rapidly in normal circumstances. Also, the fact that hydrogen has to be used as a fuel is a disadvantage. To solve these and other problems many different fuel cell types have been tried. The different types are usually distinguished by the electrolyte that is used, though there are always other important differences as well. Most of these fuel cells have

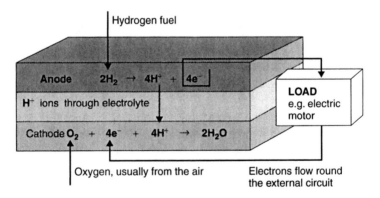

Figure 4.3 The reactions at the electrodes, and the electron movement, in a fuel cell with an acid electrolyte

Table 4.1 Data for different types of fuel cell

Fuel cell type	Mobile ion	Operating temp.	Applications and notes
Alkaline (AFC)	OH^-	50–200°C	Used in space vehicles, e.g. Apollo, Shuttle.
Proton exchange membrane (PEMFC)	H^+	30–100°C	Vehicles and mobile applications, and for lower power CHP systems
Direct methanol (DMFC)	H^+	20–90°C	Suitable for portable electronic systems of low power, running for long times
Phosphoric acid (PAFC)	H^+	~220°C	Large numbers of 200 kW CHP systems in use
Molten carbonate (MCFC)	CO_3^{2-}	~650°C	Suitable for medium to large scale CHP systems, up to MW capacity
Solid oxide (SOFC)	O^{2-}	500–1000°C	Suitable for all sizes of CHP systems, 2 kW to multi MW

somewhat different electrode reactions than those given above, however such details are given elsewhere (Larminie and Dicks 2003).

The situation now is that six classes of fuel cell have emerged as viable systems for the present and near future. Basic information about these systems is given in Table 4.1.

As well as facing up to different problems, the various fuel types also try to play to the strengths of fuel cells in different ways. The PEM fuel cell capitalises on the essential simplicity of the fuel cell. The electrolyte is a solid polymer, in which protons are mobile. The chemistry is the same as the acid electrolyte fuel cell of Figure 4.3 above. With a solid and immobile electrolyte, this type of cell is inherently simple; it is the type that shows by far the most promise for vehicles, and is the type used on all the most impressive demonstration fuel cell vehicles. This type of fuel cell is the main focus of this chapter.

PEM fuel cells run at quite low temperatures, so the problem of slow reaction rates has to be addressed by using sophisticated catalysts and electrodes. Platinum is the catalyst, but developments in recent years mean that only minute amounts are used, and the cost of the platinum is a small part of the total price of a PEM fuel cell. The problem of hydrogen supply is not really addressed; quite pure hydrogen must be used, though various ways of supplying this are possible, as is discussed in Chapter 5.

One theoretically very attractive solution to the hydrogen supply problem is to use methanol[1] as a fuel instead. This can be done in the PEM fuel cell, and such cells are called direct methanol fuel cells. 'Direct' because they use the methanol as the fuel as it is, in liquid form, as opposed to extracting the hydrogen from the methanol using one of the methods described in Chapter 5. Unfortunately these cells have very low power, and for the foreseeable future at least their use will be restricted to applications requiring slow and steady generation of electricity over long periods. A demonstration DMFC powered go-kart has been built, but really the only likely application of this type of cell in the near future is in the rapidly growing area of portable electronics equipment.

---

[1] A fairly readily available liquid fuel, formula CH_3OH.

Although PEM fuel cells were used on the first manned spacecraft, the alkaline fuel cell was used on the Apollo and is used on the Shuttle Orbiter. The problem of slow reaction rate is overcome by using highly porous electrodes, with a platinum catalyst, and sometimes by operating at quite high pressures. Although some historically important alkaline fuel cells have operated at about 200°C, they more usually operate below 100°C. The alkaline fuel cell has been used by a few demonstration electric vehicles, always in hybrid systems with a battery. They can be made more cheaply than PEMFCs, but they are lower in power, and the electrolyte reacts with carbon dioxide in the air, which make terrestrial applications difficult.

The phosphoric acid fuel cell (PAFC) was the first to be produced in commercial quantity and enjoy widespread terrestrial use. Many 200 kW systems, manufactured by the International Fuel Cells Corporation, are installed in the USA and Europe, as well as systems produced by Japanese companies. However, they are not suitable for vehicles, as they operate at about 220°C, and do not react well to being cooled down and re-started; they are suited to applications requiring power all the time, day after day, month after month.

As is the way of things, each fuel cell type solves some problems, but brings new difficulties of its own. The solid oxide fuel cell (SOFC) operates in the region of 600 to 1000°C. This means that high reaction rates can be achieved without expensive catalysts, and that gases such as natural gas can be used directly, or 'internally reformed' within the fuel cell; they do not have to have a hydrogen supply. This fuel cell type thus addresses some of the problems and takes full advantage of the inherent simplicity of the fuel cell concept. Nevertheless, the ceramic materials that these cells are made from are difficult to handle, so they are expensive to manufacture, and there is still quite a large amount of extra equipment needed to make a full fuel cell system. This extra plant includes air and fuel pre-heaters, also the cooling system is more complex, and they are not easy to start up. No-one is developing these fuel cells as the motive power unit for vehicles, but some are developing smaller units to provide the electric power for air conditioning and other systems on modern 'conventional' engined vehicles, which have very high electric power demands these days. However, that is not the focus of this book.

Despite operating at temperatures of up to 1000°C, the SOFC always stays in the solid state. This is not true for the molten carbonate fuel cell (MCFC), which has the interesting feature that it needs the carbon dioxide in the air to work. The high temperature means that a good reaction rate is achieved using a comparatively inexpensive catalyst, nickel. The nickel also forms the electrical basis of the electrode. Like the SOFC it can use gases such as methane and coal gas (H_2 and CO) as fuel. However, this simplicity is somewhat offset by the nature of the electrolyte, a hot and corrosive mixture of lithium, potassium and sodium carbonates. They are not suitable for vehicles,[2] as they only work well as rather large systems, running all the time.

So, fuel cells are very varied devices, and have applications way beyond vehicles. For the rest of this chapter we will restrict ourselves to the PEM fuel cell, as it is by far the most important in this context.

---

[2] Except ships.

4.2.3 Fuel cell electrodes

Figure 4.4 is another representation of a fuel cell. Hydrogen is fed to one electrode, and oxygen, usually as air, to the other. A load is connected between the two electrodes, and current flows. However, in practice a fuel cell is far more complex than this. Normally the rate of reaction of both hydrogen and oxygen is very slow, which results in a low current, and so a low power. The three main ways of dealing with the slow reaction rates are: the use of suitable catalysts on the electrode, raising the temperature, and increasing the electrode area.

The first two can be applied to any chemical reaction. However, the third is special to fuel cells and is very important. If we take a reaction such as that of equation (4.3), we see that oxygen gas, and H^+ ions from the electrolyte, and electrons from the circuit are needed, all three together. This 'coming together' must take place on the surface of the electrode. Clearly, the larger the electrode area, the more scope there is for this to happen, and the greater the current. This is very important. Indeed, electrode area is such a vital issue that the performance of a fuel cell design is often quoted in terms of the current *per cm²*.

The structure of the electrode is also important. It is made highly porous so that the real surface area is much greater than the normal length × width.

As well as being of a large surface area, and highly porous, a fuel cell electrode must also be coated with a catalyst layer. In the case of the PEMFC this is platinum, which

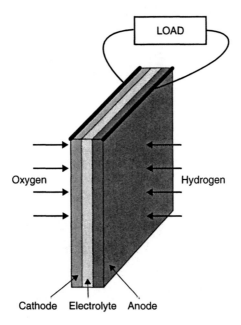

Figure 4.4 Basic cathode-electrolyte-anode construction of a fuel cell. Note that the anode is the negative terminal, and the cathode the positive. This may seem counter to expectations, but is in fact true for all primary cell. The rule is that the cathode is the terminal that the electrons flow *into*. So, in electrolysis cells the cathode is the negative

is highly expensive. The catalyst thus needs to be spread out as finely as possible. This is normally done by supporting very fine particles of the catalyst on carbon particles. Such a carbon-supported catalyst is shown for real in Figure 4.5, and in idealised form in Figure 4.6.

The reactants need to be brought into contact with the catalyst, and a good electrical contact needs to be made with the electrode surface. Also, in the case of the cathode, the product water needs to be removed. These tasks are performed by the 'gas diffusion layer', a porous and highly conductive material such as carbon felt or carbon paper, which is layered on the electrode surface.

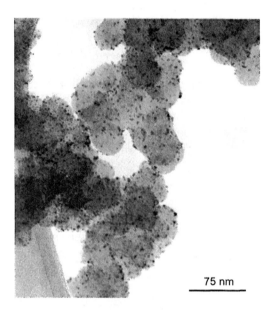

Figure 4.5 Electron microscope image of some fuel cell catalyst. The black specks are the catalyst particles finely divided over larger carbon supporting particles (Reproduced by kind permission of Johnson Matthey plc.)

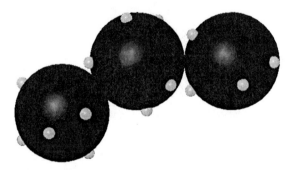

Figure 4.6 The structure, idealised, of carbon-supported catalyst

Fuel Cells

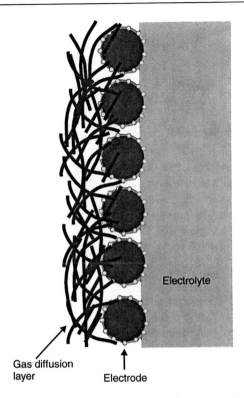

Figure 4.7 Simplified and idealised structure of a PEM fuel cell electrode

Finally, some of the electrode is allowed to permeate over the surface of the carbon supported catalyst to increase the contact between reactants. The resulting structure is shown, in somewhat idealised form, in Figure 4.7. All items shown in this diagram are in reality very thin. The electrolyte is about 0.05 to 0.1 mm thick, and each electrode is about 0.03 mm thick, with the gas diffusion layers each about 0.2 to 0.5 mm thick. The whole anode/electrolyte/cathode assembly, often called a membrane electrode assembly or MEA, is thus typically about 1 mm thick, including the gas diffusion layers.

4.3 Fuel Cell Thermodynamics – an Introduction

4.3.1 Fuel cell efficiency and efficiency limits

One of the attractions of fuel cells is that they are not heat engines. Their thermodynamics are different, and in particular their efficiency is potentially greater as they are not limited by the well-known Carnot limit that impinges on IC and other types of fuel burning engines. However, as we shall see, they do have their own limitations, and while fuel cells are often more efficient than IC engines, the difference is sometimes exaggerated.

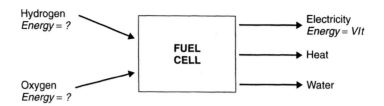

Figure 4.8 Fuel cell inputs and outputs

At first we must acknowledge that the efficiency of a fuel cell is not straightforward to define. In some electrical power generating devices it is very clear what form of energy is being converted into electricity. With a fuel cell such energy considerations are much more difficult to visualise. The basic operation has already been explained, and the input and outputs are shown in Figure 4.8. The electrical power and energy output are easily calculated from the well known formulas:

$$\text{Power} = VI \quad \text{and} \quad \text{Energy} = VIt$$

However, the energy of the chemical inputs and output is not so easily defined. At a simple level we could say that it is the chemical energy of the H_2, O_2 and H_2O that is in question. The problem is that chemical energy is not simply defined, and terms such as enthalpy, Helmholtz function and Gibbs free energy are used. In recent years the useful term 'exergy' has become quite widely used, and the concept is particularly useful in high temperature fuel cells, though we are not concerned with these here. There are also older (but still useful) terms such as calorific value.

In the case of fuel cells it is the Gibbs free energy that is important. This can be defined as the energy available to do external work, neglecting any work done by changes in pressure and/or volume. In a fuel cell the external work involves moving electrons round an external circuit; any work done by a change in volume between the input and output is not harnessed by the fuel cell.[3] Exergy is *all* the external work that can be extracted, including that due to volume and pressure changes. Enthalpy, simply put, is the Gibbs free energy plus the energy connected with the entropy. The enthalpy H, Gibbs free energy G and entropy S are connected by the well-known equation:

$$G = H - TS$$

The energy that is released by a fuel cell is the change in Gibbs energy before and after a reaction, so the energy released can be represented by the equation:

$$\Delta G = G_{outputs} - G_{inputs}$$

However, the Gibbs free energy change is *not constant*, but changes with temperature and state (liquid or gas). Table 4.2 below shows ΔG for the basic hydrogen fuel cell reaction

$$H_2 + \tfrac{1}{2}O_2 \longrightarrow H_2O$$

---

[3] It may be harnessed by some kind of turbine in a combined cycle system, as discussed in Chapter 6.

Fuel Cells

Table 4.2 ΔG for the reaction $H_2 + \frac{1}{2}O_2 \rightarrow H_2O$ at various temperatures

Form of water product	Temperature (°C)	ΔG (kJ/mole)
Liquid	25	−237.2
Liquid	80	−228.2
Gas	80	−226.1
Gas	100	−225.2
Gas	200	−220.4
Gas	400	−210.3
Gas	600	−199.6
Gas	800	−188.6
Gas	1000	−177.4

for a number of different conditions. Note that the values are negative, which means that energy is released.

If there are no losses in the fuel cell, or as we should more properly say, if the process is reversible, then all this Gibbs free energy is converted into electrical energy. We could thus define the efficiency of a fuel cell as:

$$\frac{\text{electrical energy produced}}{\text{Gibbs free energy change}}$$

However, this is not very useful, and is rarely done, not least because the Gibbs free energy change is not constant.

Since a fuel cell uses materials that are usually burnt to release their energy, it would make sense to compare the electrical energy produced with the heat that would be produced by burning the fuel. This is sometimes called the calorific value, though a more precise description is the change in enthalpy of formation. Its symbol is ΔH. As with the Gibbs free energy, the convention is that ΔH is negative when energy is released. So to get a good comparison with other fuel using technologies, the efficiency of the fuel cell is usually defined as:

$$\frac{\text{electrical energy produced per mole of fuel}}{-\Delta H} \quad (4.4)$$

However, even this is not without its ambiguities, as there are two different values that we can use for ΔH. For the burning of hydrogen:

$$H_2 + \tfrac{1}{2}O_2 \longrightarrow H_2O \text{ (steam)}$$
$$\Delta H = -241.83 \text{ kJ/mole}$$

Whereas if the product water is condensed back to liquid, the reaction is:

$$H_2 + \tfrac{1}{2}O_2 \longrightarrow H_2O \text{ (liquid)}$$
$$\Delta H = -285.84 \text{ kJ/mole}$$

The difference between these two values for ΔH (44.01 kJ/mole) is the molar enthalpy of vaporisation[4] of water. The higher figure is called the higher heating value (HHV), and the lower, quite logically, the lower heating value (LHV). Any statement of efficiency should say whether it relates to the higher or lower heating value. If this information is not given, the LHV has probably been used, since this will give a higher efficiency figure.

We can now see that there is a limit to the efficiency, if we define it as in equation (4.4). The maximum electrical energy available is equal to the change in Gibbs free energy, so:

$$\text{Maximum efficiency possible} = \frac{\Delta G}{\Delta H} \times 100\% \tag{4.5}$$

This maximum efficiency limit is sometimes known as the thermodynamic efficiency. Table 4.3 gives the values of the efficiency limit, relative to the higher heating value, for a hydrogen fuel cell. The maximum voltage obtainable from a single cell is also given.

The graphs in Figure 4.9 show how these values vary with temperature, and how they compare with the Carnot limit, which is given by the equation:

$$\text{Carnot limit} = \frac{T_1 - T_2}{T_1}$$

where T_1 is the higher temperature, and T_2 the lower, of the heat engine. The graph makes clear that the efficiency limit of the fuel cell is certainly not 100%, as some supporters of fuel cells occasionally claim. Indeed, above the 750°C the efficiency limit of the hydrogen fuel cell is actually less than for a heat engine. Nevertheless, the PEM fuel cells used in vehicles operate at about 80°C, and so their theoretical maximum efficiency is actually much better than for an IC engine.

4.3.2 Efficiency and the fuel cell voltage

A very useful feature of fuel cells is that their efficiency can be very easily found from their operating voltage. The reasoning behind this is as follows. If *one* mole of fuel is reacted in the cell, then *two* moles of electrons are pushed round the external circuit; this

Table 4.3 ΔG, maximum EMF, and efficiency limit (HHV) for hydrogen fuel cells

Form of water product	Temp °C	ΔG kJ/mole^{-1}	Max. EMF	Efficiency limit
Liquid	25	−237.2	1.23 V	83%
Liquid	80	−228.2	1.18 V	80%
Gas	100	−225.3	1.17 V	79%
Gas	200	−220.4	1.14 V	77%
Gas	400	−210.3	1.09 V	74%
Gas	600	−199.6	1.04 V	70%
Gas	800	−188.6	0.98 V	66%
Gas	1000	−177.4	0.92 V	62%

---

[4] This used to be known as the molar 'latent heat'.

Fuel Cells

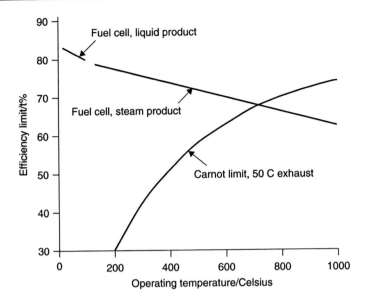

Figure 4.9 Maximum hydrogen fuel cell efficiency at standard pressure, with reference to the higher heating value. The Carnot limit is shown for comparison, with a 50°C exhaust temperature

can be deduced from Figure 4.3. We also know that the electrical energy is given by the fundamental energy equation:

$$\text{Energy} = \text{Charge} \times \text{Voltage}$$

The Faraday constant F gives the charge on one mole of electrons. So, when one mole of hydrogen fuel is used in a fuel cell, if it were 100% efficient, as defined by equation (4.4), then we would be able to say that:

$$\text{Energy} = 2F \times V_{100\%} = \Delta H$$

$$\text{and thus } V_{100\%} = \frac{\Delta H}{2F}$$

Using standard values for the Faraday constant (96 485 Coulombs), and the two values for ΔH given above, we can easily calculate that the '100% efficient' voltage for a single cell is 1.48 V if using the HHV or 1.25 V if using the LHV.

Now of course a fuel cell never is, and we have shown in the last section never can be, 100% efficient. The actual fuel cell voltage will be a lower value, which we can call V_o. Since voltage and electrical energy are directly proportional, it is clear that

$$\text{Fuel cell efficiency} = \frac{V_c}{V_{100\%}} = \frac{V_c}{1.48} \tag{4.6}$$

Clearly it is very easy to measure the voltage of a fuel cell. In the case of a stack of many cells, remember that the voltage of concern is the average voltage **one** cell, so the system voltage should be divided by the number of cells. The efficiency can thus be found remarkably easily.

It is worth noting in passing that the maximum voltage of a fuel cell occurs when 100% of the Gibbs free energy is converted into electrical energy. Thus we have a 'sister' equation to equation (4.4), giving the maximum possible fuel cell voltage:

$$V_{max} = \frac{\Delta G}{2F} \qquad (4.7)$$

This is also a very important fuel cell equation, and it was used to find the figures shown in the fourth column of Table 4.3.

4.3.3 Practical fuel cell voltages

Equation (4.7) above gives the maximum possible voltage obtainable from a single fuel cell. In practice the actual cell voltage is less than this. Now of course this applies to ordinary batteries too, as when current is drawn out of any electric cell the voltage falls, due to internal resistances. However, with a fuel cell this effect is more marked than with almost all types of conventional cell. Figure 4.10 shows a typical voltage/current density curve for a good PEM fuel cell. It can be seen that the voltage is always less, and is often much less, than the 1.18 V that would be obtained if all of the Gibbs energy were converted into electrical energy.

There are three main reasons for this loss of voltage, as detailed below.

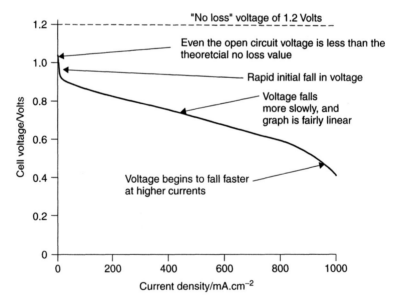

Figure 4.10 Graph showing the voltage from a typical good quality PEM fuel cell operating on air at about 80°C

- The energy required to drive the reactions at the electrodes, usually called the activation energy, causes a voltage drop. This is especially a problem at the air cathode, and shows itself as a fairly constant voltage drop. This explains the initial fall in voltage even at quite low currents.
- The resistance of the electrolyte and the electrodes causes a voltage drop that more-or-less follows Ohm's law, and causes the steady fall in voltage over the range of currents. This is usually called the Ohmic voltage loss.
- At very high currents, the air gets depleted of oxygen, and the remnant nitrogen gets in the way of supplying fresh oxygen. This results is a fall in voltage, as the electrodes are short of reactant. This problem causes the more rapid fall in voltage at higher currents, and is called mass transfer or concentration voltage loss.

A result of the huge effort in fuel cell development over the last ten years or so has resulted in great improvement in the performance of fuel cells, and a reduction in all these voltage losses. A fuel cell will typically operate at an average cell voltage of about 0.65 to 0.70 V, even at currents approaching 1 A per cm^2. This represents an efficiency of about 50% (with respect to the HHV), which is considerably better than any IC engine, though some of the electrical energy is used up driving the fuel cell ancillary equipment to be discussed in the sections that follow.

We should point out that a consequence of the higher cell voltage at lower currents is that the efficiency is higher at lower currents. This is a marked contrast to the IC engine, where the efficiency is particularly poor at low powers.

In the opening section of this chapter we pointed out that a fuel cell could be compared to an IC engine running on hydrogen fuel, which would also give out very limited pollution. This section has shown that fuel cells do have the potential to have a considerably higher efficiency than IC engines, and so they would, all other things being equal, be preferred. The problem is that all other things are not equal. At the moment fuel cells are vastly more expensive than IC engines, and this may remain so for some time. It is by no means clear-cut that fuel cells are the better option. A hydrogen powered IC engine in a hybrid drive train would be not far behind a fuel cell in efficiency, and the advantages of proven and available technology might tip the balance against higher efficiency and even less pollution. Time will tell.

4.3.4 The effect of pressure and gas concentration

The values for the changes in the Gibbs free energy given in Tables 4.2 and 4.3 all concern pure hydrogen and oxygen, at standard pressure, 100 kPa. However, as well as changing with temperature, as shown in these tables, the Gibbs energy changes with pressure and concentration.

A full treatment of these issues is beyond a book such as this, and it can easily be found elsewhere (e.g. Chapter 2 of Larminie and Dicks 2003). Suffice to say that the relationship is given by a very important fuel cell equation derived from the work of Nernst. It can be expressed in many different forms, depending on what issue is to be analysed. For example, if the change of system pressure is the issue, then the Nernst equation takes the form:

$$\Delta V = \frac{RT}{4F} \ln\left(\frac{P_2}{P_1}\right) \quad (4.8)$$

Where ΔV is the voltage increase if the pressure changes from P_1 to P_2. Other causes of voltage change are a reduction in voltage caused by using air instead of pure oxygen. The use of hydrogen fuel that is mixed with carbon dioxide, as is obtained from the 'reforming' of fuels such as petrol, methanol or methane (as described in Chapter 5), also causes a small reduction in voltage.

For high temperature fuel cells the Nernst equation predicts very well the voltage changes. However, with lower temperature cells, such as are used in electric vehicles, the changes are nearly always considerably greater than the Nernst equation predicts. This is because the 'activation voltage drop' mentioned in the last section is also quite strongly affected by issues such as gas concentration and pressure. This is especially the case at the air cathode.

For example, equation (4.8) would predict that for a PEM fuel cell working at 80°C, the voltage increase resulting from a doubling of the system pressure would be:

$$\Delta V = \frac{8.314 \times (273 + 80)}{4 \times 96485} \ln(2) = 0.0053 \text{ V per cell}$$

However, in practice the voltage increase would typically be about 0.04 V, nearly ten times as much. Even so, we should note that the increase is still not large, and that there is considerable energy cost in running the system at higher pressure. Indeed, it is shown elsewhere (e.g. Larminie and Dicks 2003, Chapter 4) that the energy gained from a higher voltage is very unlikely to be greater than the energy loss in pumping the air to higher pressure.

Nevertheless, it is the case that most PEM fuel cells in vehicle applications *are* run at a pressure distinctly above air pressure, typically between 1.5 and 2.0 bar. The reasons for this are not primarily because of increasing the cell voltage. Rather, they are because it makes the water balance in the PEM fuel cell much easier to maintain. This complicated issue is explained in Section 4.5 below.

4.4 Connecting Cells in Series – the Bipolar Plate

As has been pointed out in the previous section, the voltage of a working fuel cell is quite small, typically about 0.7 V when drawing a useful current. This means that to produce a useful voltage many cells have to be connected in series. Such a collection of fuel cells in series is known as a 'stack'. The most obvious way to do this is by simply connecting the edge of each anode to the cathode of the next cell all along the line, as in Figure 4.11. (For simplicity, this diagram ignores the problem of supplying gas to the electrodes.)

The problem with this method is that the electrons have to flow across the face of the electrode to the current collection point at the edge. The electrodes might be quite good conductors, but if each cell is only operating at about 0.7 V, even a small voltage drop is important. Unless the current flows are very low, and the electrode a particularly good conductor, or very small, this method is not used.

The method of connecting to a single cell, all over the electrode surfaces, while at the same time feeding hydrogen to the anode and oxygen to the cathode, is shown in Figure 4.12. The grooved plates are made of a good conductor such as graphite or stainless

Fuel Cells

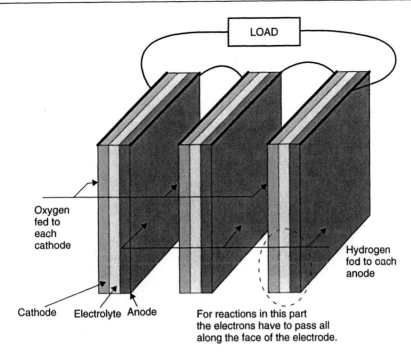

Figure 4.11 Simple edge connection of three cells in series

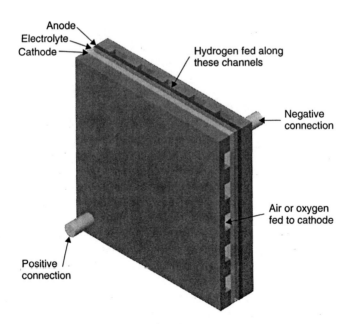

Figure 4.12 Single cell, with end plates for taking the current from all over the face of the electrodes, and also supplying gas to the whole electrode

Figure 4.13 Two bipolar plates of very simple design. There are horizontal grooves on one side and vertical grooves on the other

steel. This idea is then extended to the 'bipolar plate' shown in Figure 4.13. These make connections all over the surface of one cathode and also the anode of the next cell (hence 'bipolar'). At the same time the bipolar plate serves as a means of feeding oxygen to the cathode and hydrogen to the anode. A good electrical connection must be made between the two electrodes, but the two gas supplies must be strictly separated, otherwise a dangerous hydrogen/oxygen mixture will be produced.

However, this simple type of bipolar plate shown in Figure 4.13 will not do for PEM fuel cells. Because the electrodes must be porous (to allow the gas in) they would allow the gas to leak out of their edges. The result is that the edges of the electrodes must be sealed. This is done by making the electrolyte somewhat larger than the electrodes, and fitting a sealing gasket around each electrode, as shown in Figure 4.14.

Rather than feeding the gas in at the edge, as in Figures 4.12 and 4.13, a system of 'internal manifolding' is used with PEM fuel cells. This arrangement requires a more complex bipolar plate, and is shown in Figure 4.15. The plates are made larger relative to the electrodes, and have extra channels running through the stack which feed the fuel and oxygen to the electrodes. Carefully placed holes feed the reactants into the channels that run over the surface of the electrodes. It results in a fuel cell stack that has the appearance of the solid block, with the reactant gases fed in at the ends, where the positive and negative connections are also made.

Figure 4.16 shows a fairly high power PEM fuel cell system. It consists of four stacks made as described above, each a block of approximately square cross-section.

Fuel Cells

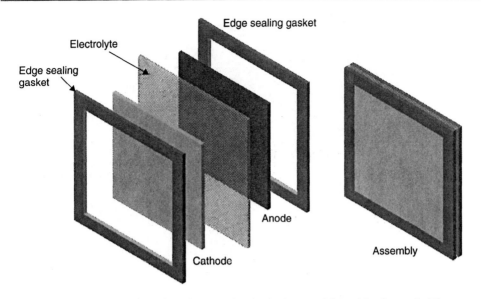

Figure 4.14 The construction of anode/electrolyte/cathode assemblies with edge seals. These prevent the gases leaking out of the edge of the porous electrodes

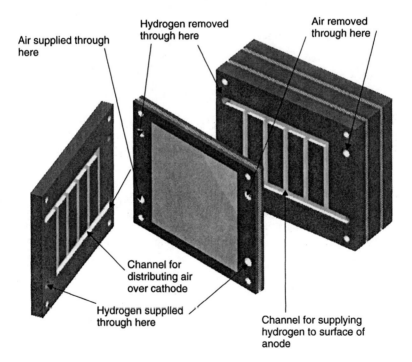

Figure 4.15 A simple bipolar plate with internal manifolding, as is usually used in PEM fuel cells. The reactant gases are fed to the electrodes through internal tubes

Figure 4.16 An example of a fairly large PEM fuel cell system. Four separate fuel cell stacks can be seen. Each stack consists of 160 cells in series (Reproduced by kind permission of MAN Nutzfahrzeuge AG. This is the stack from the bus in Figure 4.2, and is made by Siemens.)

A further complication is that the bipolar plates also have to incorporate channels in them for cooling water or air to pass through, as fuel cells are not 100% efficient and generate heat as well as electricity.

It should now be clear that the bipolar plate is quite a complex item. A fuel cell stack, such as those of Figure 4.16, will have up to 80 cells in series, and so a large number will be needed. As well as being a fairly complex item to make, the question of its material is often difficult. Graphite, for example, can be used, but this is difficult to work and is brittle. Stainless steel can also be used, but this will corrode in some types of fuel cell. To form the gas flow paths, and to make the plates quickly and cheaply, plastic would be ideal. However, the bipolar plate must clearly be a very good conductor of electricity, and this is a great difficulty for plastics. The present situation is that no entirely satisfactory way of making these items has yet been developed, but many of the most promising options are discussed elsewhere, such as Ruge and Büchi (2001). It is certainly the case now, and will be for many years, that the bipolar plate makes a major contribution to the cost of a fuel cell, as well as its size and its weight.

Anyone who has made fuel cells knows that leaks are a major problem. If the path of hydrogen through a stack using internal manifolding (as in Figure 4.15) is imagined, the possibilities for the gas to escape are many. The gas must reach the edge of every porous electrode, so the entire edge of every electrode is a possible escape route, both under and over the edge gasket. Other likely trouble spots are the joins between each and every bipolar plate. In addition, if there is the smallest hole in any of the electrolyte, a serious leak is certain.

The result is that a fuel cell is quite a difficult system to manufacture, requiring parts that are complex to form rapidly and cheaply. Very careful assembly is required, and each fuel cell stack consists of a large number of components. The system has a very low level of fault tolerance.

4.5 Water Management in the PEM Fuel Cell

4.5.1 Introduction to the water problem

We see in Figure 4.3 the different electrode reactions in a fuel cell. Looking back at this diagram, you will see that the water product from the chemical reaction is made on the positive electrode, where air is supplied.

This is highly convenient. It means that air can be supplied to this electrode, and as it blows past it will supply the necessary oxygen, and also evaporate off the product water and carry it off, out of the fuel cell.

This is indeed what happens, in principle, in the PEM fuel cell. However, unfortunately the details are far more complex and much more difficult to manage. The reasons for this require that we understand in some detail the operation of the electrolyte of a PEM fuel cell.

4.5.2 The electrolyte of a PEM fuel cell

The different companies producing polymer electrolyte membranes have their own special tricks, mostly proprietary. However, a common theme is the use of sulphonated fluoropolymers, usually fluoroethylene. The most well known and well established of these is Nafion (® Dupont), which has been developed through several variants since the 1960s. This material is still the electrolyte against which others are judged, and is in a sense an 'industry standard'. Other polymer electrolytes function in a similar way.[5]

The construction of the electrolyte material is as follows. The starting point is the basic and simplest to understand man-made polymer, polyethylene. Based on ethylene, its molecular structure is shown in Figure 4.17.

This basic polymer is modified by substituting fluorine for the hydrogen. This process is applied to many other compounds, and is called 'perfluorination'. The 'mer' is

---

[5] For a review of work with other types of proton exchange membrane, see Rozière and Jones (2001).

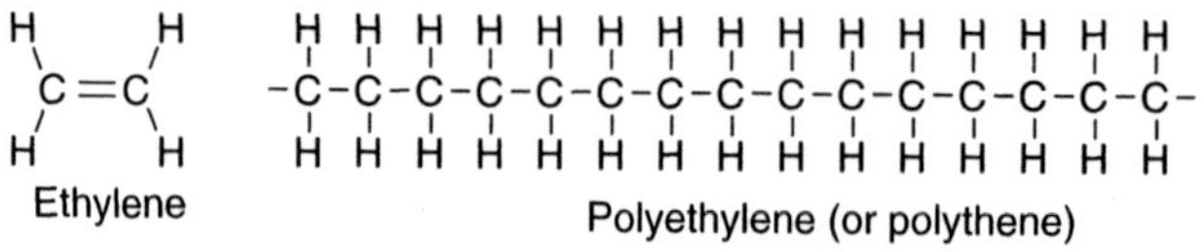

Figure 4.17 The structure of polyethylene

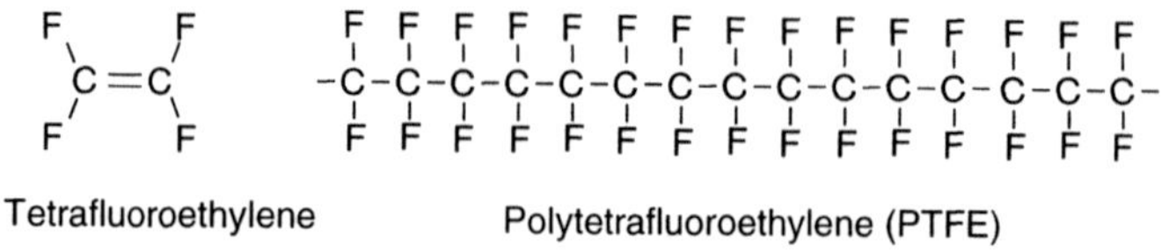

Figure 4.18 The structure of PTFE

called tetrafluoroethylene.[6] The modified polymer, shown in Figure 4.18, is polytetrafluoroethylene, or PTFE. It is also sold as Teflon, the registered trademark of ICI. This remarkable material has been very important in the development of fuel cells. The strong bonds between the fluorine and the carbon make it highly resistant to chemical attack and durable. Another important property is that it is strongly hydrophobic, and so it is used in fuel cell electrodes to drive the product water out of the electrode, and thus prevent flooding. It is used in this way in phosphoric acid and alkali fuel cells, as well as PEMFCs. (The same property gives it a host of uses in outdoor clothing and footwear.)

However, to make an electrolyte, a further stage is needed. The basic PTFE polymer is 'sulphonated'; a side chain is added, ending with sulphonic acid HSO_3. Sulphonation of complex molecules is a widely used technique in chemical processing. It is used, for example, in the manufacture of detergent. One possible side chain structure is shown in Figure 4.19; the details vary for different types of Nafion, and with different manufacturers of these membranes. The methods of creating and adding the side chains is proprietary, though one modern method is discussed by Kiefer *et al.* (1999).

The HSO_3 group added is ionically bonded, and so the end of the side chain is actually an SO_3^- ion. The result of the presence of these SO_3^- and H^+ ions is that there is a strong mutual attraction between the + and − ions from each molecule. The result is that the side chain molecules tend to cluster within the overall structure of the material. Now, a key property of sulphonic acid is that it is highly hydrophyllic, it attracts water. (This is why it is used in detergent; it makes one end of the molecule mix readily with water, while the other end attaches to the dirt.) In Nafion, this means we are creating hydrophyllic regions within a generally hydrophobic substance, which is bound to create interesting results.

The hydrophyllic regions around the clusters of sulphonated side chains can lead to the absorption of large quantities of water, increasing the dry weight of the material by up to 50%. Within these hydrated regions the H^+ ions are relatively weakly attracted to the SO_3^- group, and are able to move. This creates what is essentially a dilute acid. The resulting material has different phases, dilute acid regions within a tough and strong hydrophobic

---

[6] 'Tetra' indicates that all four hydrogens in each ethylene group have been replaced by fluorine.

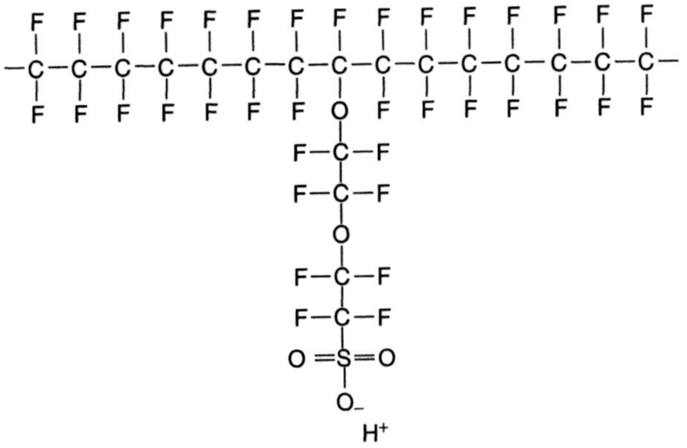

Fuel Cells

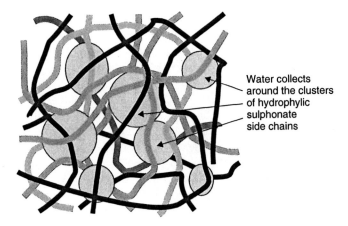

Figure 4.19 Example structure of a sulphonated fluoroethylene, also called perfluorosulphonic acid PTFE copolymer

Figure 4.20 The structure of Nafion-type membrane materials. Long chain molecules containing hydrated regions around the sulphonated side chains

structure. This is illustrated in Figure 4.20. Although the hydrated regions are somewhat separate, it is still possible for the H^+ ions to move through the supporting long molecule structure. However, it is easy to see that for this to happen the hydrated regions must be as large as possible. In a well hydrated electrolyte there will be about 20 water molecules for each SO_3^- side chain. This will typically give a conductivity of about $0.1\,\text{Scm}^{-1}$. As the water content falls, so the conductivity falls in a more or less linear fashion.

From the point of view of fuel cell use, the main features of Nafion and other fluorosulphonate ionomers are that:

- they are highly chemically resistant;
- they are mechanical strong, and so can be made into very thin films, down to $50\,\mu\text{m}$;

- they are acidic;
- they can absorb large quantities of water;
- if they are well hydrated, then H$^+$ ions can move quite freely within the material, so they are good proton conductors.

This material then is the basis of the proton exchange membrane (PEM) fuel cell. It is not cheap to manufacture, but costs could fall if production was on a really large scale. The key point to remember for the rest of this section is that for the electrolyte to work properly, it must be very well hydrated.

4.5.3 Keeping the PEM hydrated

It will be clear from the description of a proton exchange membrane given in the last section that there must be sufficient water content in the polymer electrolyte. The proton conductivity is directly proportional to the water content. However, there must not be so much water that the electrodes, which are bonded to the electrolyte, flood and block the pores in the electrodes or gas diffusion layer. A balance is therefore needed, which takes care to achieve.

In the PEMFC water forms at the cathode; revisit Figure 4.3 if you are not sure why. In an ideal world this water would keep the electrolyte at the correct level of hydration. Air would be blown over the cathode, and as well as supplying the necessary oxygen it would dry out any excess water. Because the membrane electrolyte is so thin, water would diffuse from the cathode side to the anode, and throughout the whole electrolyte a suitable state of hydration would be achieved without any special difficulty. This happy situation can sometimes be achieved, but needs good engineering design to bring to pass.

There are several complications. One is that during the operation of the cell the H$^+$ ions moving from the anode to the cathode (see Figure 4.3) pull water molecules with them. This process is sometimes called 'electro-osmotic drag'. Typically between 1 and 5 water molecules are 'dragged' for each proton (Zawodzinski *et al.* 1993, Ren and Gottesfeld 2001). This means that, especially at high current densities, the anode side of the electrolyte can become dried out, even if the cathode is well hydrated. Another major problem is that the water balance in the electrolyte must be correct throughout the cell. In practice, some parts may be just right, others too dry, and others flooded. An obvious example of this can be seen if we think about the air as it passes through the fuel cell. It may enter the cell quite dry, but by the time it has passed over some of the electrodes it may be about right. However, by the time it has reached the exit it may be so saturated that it cannot dry off any more excess water. Obviously, this is more of a problem when designing larger cells and stacks.

Yet another complication is the drying effect of air at high temperatures. If the PEM fuel cell is operating at about 85°C, then it becomes very hard not to dry out the electrolyte. Indeed, it can be shown[7] that at temperatures of over about 65°C the air will *always* dry out the electrodes faster than water is produced by the H$_2$/O$_2$ reaction. However, operation at temperatures of about 85°C or so is essential if enough power is to be extracted for automotive applications.

---

[7] Büchi and Srinivasan (1997).

The only way to solve these problems is to humidify the air, the hydrogen or both, before they enter the fuel cell. This may seem bizarre, as it effectively adds by-product to the inputs to the process, and there cannot be many other processes where this is done. However, in the larger, warmer PEM fuel cells used in vehicles this is always needed.

This adds an important complication to a PEM fuel cell system. The technology is fairly straightforward, and there are many ways in which it can be done. Some methods are very similar to the injection of fuel into the air stream of IC engines. Others are described in fuel cell texts. However, it will certainly add significantly to the system size, complexity and cost.

The water that is added to the air or hydrogen must come from the air leaving the fuel cell, so an important feature of an automotive fuel cell system will be a method of condensing out some of the water carried out by the damp air leaving the cell.

A further impact that the problem of humidifying the reactant gases has on the design of a PEM fuel cell system is the question of operating pressure. In Section 4.2.4 it was pointed out that raising the system pressure increases fuel cell performance, but only rarely does the gain in power exceed the power required to compress the reactant air. However, the problem of humidifying the reactant gases, and of preventing the electrolyte drying out, becomes much less if the cell is pressurised. The precise details of this are proved elsewhere,[8] but suffice to say here that if the air is compressed, then much less water needs to be added to raise the water vapour pressure to a point where the electrolyte remains well hydrated. Indeed there is some synergy between compressing the reactant gases and humidifying them, as compression (unless very slow) invariably results in heating. This rise in temperature both promotes the evaporation of water put into the gas stream, and the evaporation of the water cools that gas, and prevents it from entering the fuel cell too hot.

4.6 Thermal Management of the PEM Fuel Cell

It might be supposed that the cooling problem of a fuel cell would be simpler than for IC engines. Since they are more efficient, then less heat is generated, and so there is less heat to dispose of. Unfortunately however, this is not the case.

It is true that there is somewhat less heat energy produced. A fuel cell system will typically be about 40% efficient, compared to about 20% for an IC engine. However, in an IC engine a high proportion of the waste heat simply leaves the system in the exhaust gas.

With a fuel cell the oxygen depleted and somewhat damper air that leaves the cell will only be heated to about 85°C, and so will carry little energy. In addition, compared to an IC engine, the external surface is considerably cooler, and so far less heat is radiated and conducted away through that route.

The result is that the cooling system has to remove at least as much heat as with an IC engine, and usually considerably more.

In very small fuel cells the waste heat can be removed by passing excess air over the air cathode. This air then supplies oxygen, carries away the product water, and cools the

---

[8] Larminie and Dicks (2003), Chapter 4.

cell. However, it can be shown that this is only possible with fuel cells of power up to about 100 W. At higher powers the airflow needed is too great and far too much water would be evaporated, and the electrolyte would cease to work properly, for the reasons outlined in the previous section. Such small fuel cells have possible uses with portable electronics equipment, but are not applicable to electric vehicles.

The next stage is to have two air flows through the fuel cell. One is the 'reactant air' flowing over the fuel cell cathodes. This will typically be at about twice the rate needed to supply oxygen, so it never becomes too oxygen-depleted, but does not dry out the cell too much. The second will be the 'cooling air'. This will typically blow through channels in the bipolar plates, as shown in Figure 4.21.

This arrangement works satisfactorily in fuel cell of power up to 2 or 3 kW. Such fuel cells might one day find use in electric scooters. However, for the higher power cells to be used in cars and buses it is too difficult to ensure the necessary even air flow through the system. In this case a cooling fluid needs to be used. Water is the most common, as it has good cooling characteristics, is cheap, and the bipolar plates have in any case to be made of a material that is corrosion-resistant.

The extra cooling channels for the water (or air) are usually introduced into the bipolar plate by making it in two halves. The gas flow channels shown in Figure 4.21 are made

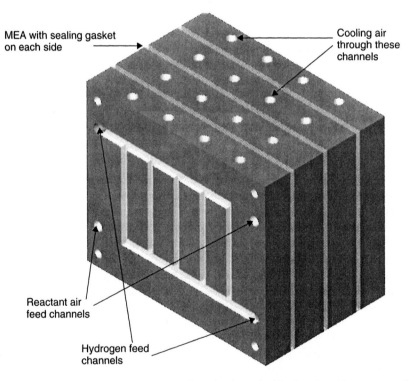

Figure 4.21 Three cells from a PEM fuel cell stack where the bipolar plated incorporate channels for cooling air, in addition to channels for reactant air over the electrodes

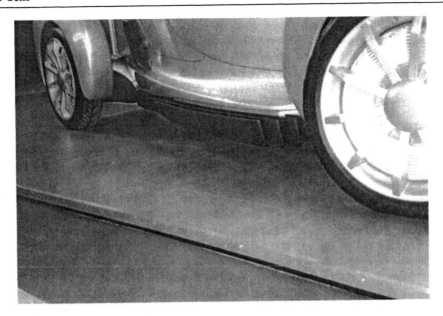

Figure 4.22 Solid metal cooling fins on the side of a GM Hy-wire demonstration fuel cell vehicle

on one face, with the cooling water channels on the other. The two halves are then joined together, giving cooling fluid channels running through the middle of the completed bipolar plate. The cooling water will then need to be pumped through a conventional heat exchanger or 'radiator', as with an IC engine. The only difference is that we will need to dispose of about twice as much heat as that of the equivalent size of IC engine.

Because larger 'radiators' are sometimes needed for fuel cells, some imagination is sometimes needed in their design and positioning. In the ground-breaking General Motors Hy-wire design, which can also be seen in Figure 4.1, large cooling fins are added to the side of the vehicle,[9] as shown in Figure 4.22.

4.7 A Complete Fuel Cell System

In Section 4.2 we explained how a fuel cell worked. We saw that, in essence, it is very simple. Hydrogen is supplied to one electrode, oxygen to the other, and electricity is produced. Pure water is the only by-product. However, in the following sections we went on to show that in practice a fuel cell is a complex system. They are difficult to make. The water balance and temperature require careful control. They consist of much more than just electrodes and electrolyte. These 'extras' are sometimes called the balance of plant (BOP).

On all but the smallest fuel cells the air and fuel will need to be circulated through the stack using *pumps* or *blowers*. In vehicles *compressors* will be used, which will be

---

[9] Note that this vehicle can also be seen in Figure 8.16, but this is a somewhat different version, which has more conventional cooling arrangements.

linked with the *humidification system* (Section 4.5). To keep this working properly, there will need to be a *water recovery* system. A *cooling system* will be needed (Section 4.6).

The DC output of a fuel cell stack will rarely be suitable for direct connection to an electrical load, and so some kind of *power conditioning* is nearly always needed. This may be as simple as a voltage regulator, or a *DC/DC converter*.[10] *Electric motors* too will nearly always be a vital part of a fuel cell system, driving the pumps, blowers and compressors mentioned above.

Various *control valves* will usually be needed, as well as *pressure regulators*. An electronic *controller* will be needed to co-ordinate the parts of the system. A special problem the controller has to deal with is the start-up and shut-down of the fuel cell system, as this can be a complex process.

This very important idea of the 'balance of plant' is illustrated in Figure 4.23, which is the fuel cell engine from a car. It uses hydrogen fuel, and the waste heat is only used to warm the car interior. The fuel cell stacks are in the rectangular block to the left of the picture. The rest of the unit (pumps, humidifier, power electronics, compressor) takes up well over half the volume of the whole system.

The presence of all this balance of plant has important implications for the efficiency of a fuel cell system, as nearly all of it requires energy to run. Back in Section 4.3.2 we saw that the efficiency of a fuel cell rises substantially if the current falls, as it is proportional to the operating voltage. However, when the balance of plant is included, this effect is largely wiped out. The power consumed by the ancillaries does not usually fall in proportion to the current, and in some cases it is fairly constant. The result is that

Figure 4.23 The 75 kW (approx.) fuel cell system used, for example, in the Mercedes A Class shown in Figure 1.14 (Reproduced by kind permission of Ballard Power Systems.)

---

[10] Together with electric motors, these circuits are explained in Chapter 6.

over a very broad range of operating powers the efficiency of most fuel cell systems, such as that of Figure 4.23, is more-or-less constant.

One aspect of fuel cells that we have not addressed so far is the very important question of 'Where does the hydrogen come from?' This is an important and wide ranging topic, and will be explored in the next chapter.

References

Büchi F.N. and Srinivasan S. (1997) Operating proton exchange membrane fuel cells without external humidification of the reactant gases. Fundamental aspects. *Journal of the Electrochemical Society*, Vol. 144, No. 8, pp. 2767–2772.

Kiefer J., Brack H-P., Huslage J., Büchi F.N., Tsakada A., Geiger F. and Schere G.G. (1999) Radiation grafting: a versatile membrane preparation tool for fuel cell applications. *Proceedings of the European Fuel Cell Forum Portable Fuel Cells Conference*, Lucerne, pp. 227–235.

Larminie J. and Dicks A. (2003) *Fuel Cell Systems Explained*. 2nd Edn Wiley, Chichester.

Ren X. and Gottesfeld S. (2001) Electro-osmotic drag of water in a poly(perfluorosulphonic acid) membrane. *Journal of the Electrochemical Society*, Vol. 148, No. 1, pp. A87–A93.

Rozière J. and Jones D. (2001) Recent progress in membranes for medium temperature fuel cells. *Proceedings of the first European PEFC Forum* (EFCF), pp. 145–150.

Ruge M. and Büchi F.N. (2001) Bipolar elements for PE fuel cell stacks based on the mould to size process of carbon polymer mixtures. *Proceedings of the First European PEFC Forum* (EFCF), pp. 299–308.

Zawodzinski T.A., Derouin C., Radzinski S., Sherman R.J., Smith V.T., Springer T.E. and Gottesfeld S. (1993) Water uptake by and transport through Nafion 117 membranes. *Journal of the Electrochemical Society*, Vol. 140, No. 4, pp. 1041–1047.

5

Hydrogen Supply

5.1 Introduction

In the last chapter we outlined the operation of fuel cells, and explained the main engineering problems with proton exchange membrane (PEM) fuel cells. However, perhaps the most difficult problem was not addressed: how to obtain the hydrogen fuel. It should be said at this point that the question of how to supply hydrogen does not only concern fuel cell vehicles. In the last chapter we alluded to the possibility (and indeed the practice) of running internal combustion (IC) engines on hydrogen. A hydrogen powered IC engine in a hybrid electric system could also provide a system with very low pollution.

There is already a considerable infrastructure for the manufacture and supply of hydrogen. It is used in large quantities as a chemical reagent, especially for oil refining and petroleum processing. It is also produced in huge quantities for the manufacture of ammonia in the fertiliser industry. The great majority of this hydrogen is produced by steam reforming of natural gas, which is outlined below in Section 5.2.

However, when it comes to providing hydrogen on a smaller scale, to mobile systems like a vehicle, then many problems occur, to which no really satisfactory solutions have yet been found. There are many ways in which the problem could be solved, and it is as yet far from clear which will emerge as the winners. The different possibilities are shown in Figure 5.1.

In terms of infrastructure changes, the simplest method would be to adapt the current large scale hydrogen production methods to a very small scale, and have 'reformers' on board vehicles that produce hydrogen from currently standard fuels such as gasoline. This approach is also explained in Section 5.2.

One solution is to use the present production methods, and have the hydrogen produced in large central plants, or by electrolysers, and stored and transported for fuel cell use as hydrogen. If such bulk hydrogen were produced by electrolysers running off electricity produced from renewable sources, or by chemical means from biomass fuels, then this would represent a system that was 'carbon dioxide neutral', and is the future as seen by the more optimistic.[1]

---

[1] However, it has to be said that at the moment the great majority of hydrogen production involves the creation of carbon dioxide.

Electric Vehicle Technology Explained   James Larminie and John Lowry
© 2003 John Wiley & Sons, Ltd   ISBN: 0-470-85163-5

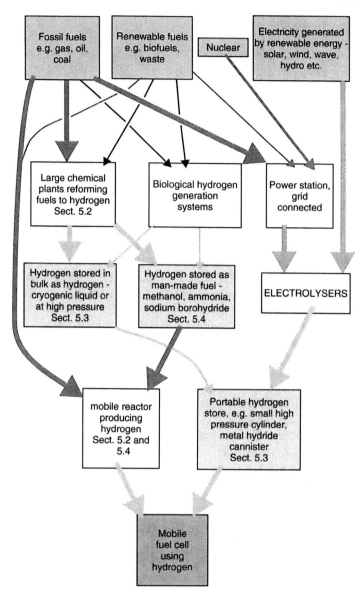

Figure 5.1 The supply of hydrogen to fuel cell powered vehicles can be achieved in many different ways

In this scenario the bulk hydrogen would be stored at local filling stations, and vehicles would 'fill up' with hydrogen, much as they do now with diesel or gasoline. Already a very few such filling stations exist, and one is shown in Figure 5.2. However, the storage of hydrogen in such stations, and even more so onboard the vehicle, is far from simple. The reasons for this are explained in Section 5.3. The problem is made more complex because

Hydrogen Supply

Figure 5.2 A hydrogen filling station. The bus in the picture is not electric, but uses a hydrogen fuelled IC engine (picture kindly supplied by MAN Nutzfahtzeuge A.G.)

some of the ways of storing hydrogen are so radically different. However, two distinct groups of methods can be identified. In one the hydrogen is stored simply as hydrogen, either compressed, or liquefied, or held in some kind of 'absorber'. The possible methods of doing this are explained in Section 5.3. This section also addresses the important issue of hydrogen safety.

In the second group of hydrogen storage methods the hydrogen is produced in large chemical plants, and is then used to produce hydrogen-rich chemicals or man-made fuels. Among these are ammonia and methanol. These 'hydrogen carrier' compounds can be made to give up their hydrogen much more easily than fossil fuels, and can be used in mobile systems. The most important of these compounds, and the ways they could be used, are explained in Section 5.4.

5.2 Fuel Reforming

5.2.1 Fuel cell requirements

Fuel reforming is the process of taking the delivered fuel, such as gasoline or propane, and converting it to a form suitable for the PEM fuel cell. This will never involve simply converting it to pure hydrogen, there will always be other substances present, particular carbon compounds.

A particular problem with fuel reformers and PEM fuel cells is the presence of carbon monoxide. This has very severe consequences for this type of fuel cell. It 'poisons' the catalyst on the electrode, and its concentration must be kept lower than about 10 parts per million. Carbon dioxide will always be present in the output of a reformer, and this poses no particular problems, except that it dilutes the fuel gas, and slightly reduces the output voltage. Steam will also be present, but as we have seen in the last chapter, this is advantageous for PEM fuel cells.

There is a very important problem that the presence of carbon dioxide in the fuel gas imposes on a fuel cell system. This is that it becomes impossible to use absolutely all of the hydrogen in the fuel cell. If the hydrogen is pure, then it can be simply connected to a fuel cell, and it will be drawn into the cell as needed. Nothing need ever come out of the fuel side of the system. When the fuel gas is impure, then it will need to be circulated through the system, with the hydrogen being used as it goes through, and with virtually 100% carbon dioxide gas at the exit. This makes for another feature of the cell that needs careful control. It also makes it important that there is still some hydrogen gas, even at the exit, otherwise the cells near the exit of the fuel flow path will not work well, as the hydrogen will be too dilute. This means that the systems described in this section will never have 100% fuel utilisation, some of it will always have to pass straight through the fuel cell stack.

5.2.2 Steam reforming

Steam reforming is a mature technology, practised industrially on a large scale for hydrogen production. The basic reforming reactions for methane and octane C_8H_{18} are:

$$CH_4 + H_2O \longrightarrow CO + 3H_2 \qquad [\Delta H = 206 \text{ kJ mol}^{-1}] \qquad (5.1)$$

$$C_8H_{18} + 8H_2O \longrightarrow 8CO + 17H_2 \qquad (5.2)$$

$$CO + H_2O \longrightarrow CO_2 + H_2 \qquad [\Delta H = -41 \text{ kJ mol}^{-1}] \qquad (5.3)$$

The reforming reactions (5.1) and (5.2), and the associated 'water-gas shift reaction' (5.3) are carried out normally over a supported nickel catalyst at elevated temperatures, typically above 500°C. Over a catalyst that is active for reactions (5.1) or (5.2), reaction (5.3) nearly always occurs as well. The combination of the two reactions taking place means that the overall product gas is a mixture of carbon monoxide, carbon dioxide and hydrogen, together with unconverted fuel and steam. The actual composition of the product from the reformer is then governed by the temperature of the reactor (actually the outlet temperature), the operating pressure, the composition of the fuel, and the proportion of steam fed to the reactor. Graphs and computer models using thermodynamic data are available to determine the composition of the equilibrium product gas for different operating conditions. Figure 5.3 is an example, showing the composition of the output at 1 bar, with methane as the fuel.

It can be seen that in the case of reaction (5.1), three molecules of carbon monoxide and one molecule of hydrogen are produced for every molecule of methane reacted. Le Chatelier's principle therefore tells us that the equilibrium will be moved to the right (i.e. in favour of hydrogen) if the pressure in the reactor is kept low. Increasing the pressure will favour formation of methane, since moving to the left of the equilibrium reduces the number of molecules.

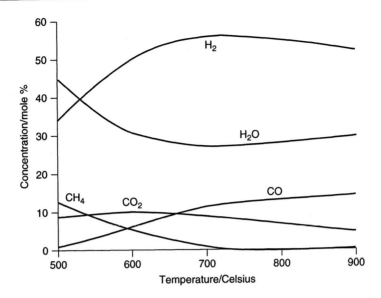

Figure 5.3 Equilibrium concentration of steam reformation reactant gases as a function of temperature. Note that at the temperature for optimum hydrogen production, considerably quantities of carbon monoxide are also produced

Another feature of reactions (5.1) and (5.2) is that they are usually *endothermic* which means that heat needs to be supplied to the reaction to drive it forward to produce hydrogen and carbon monoxide. Higher temperatures (up to 700°C) therefore favour hydrogen formation, as shown in Figure 5.3.

It is important to note at this stage that although the shift reaction (5.3) does occur at the same time as steam reforming, at the high temperatures needed for hydrogen generation, the equilibrium point for the reaction is well to the left of the equation. The result is that by no means all the carbon monoxide will be converted to carbon dioxide. For fuel cell systems that require low levels of CO, further processing will be required. These reactions are the basis of the great majority of industrial hydrogen production, using natural gas (mainly methane) as the fuel.

Hydrocarbons such as methane are not the only fuels suitable for steam reforming. Alcohols will also react in a steam reforming reaction, for example methanol:

$$CH_3OH + H_2O \longrightarrow 3H_2 + CO_2 \quad [\Delta H = 49.7\,\text{kJ}\,\text{mol}^{-1}] \quad (5.4)$$

The mildly endothermic steam reforming of methanol is one of the reasons why methanol is finding favour with vehicle manufacturers as a possible fuel for fuel cell vehicles, a point which is considered further in Section 5.4.2 below. Little heat needs to be supplied to sustain the reaction, which will readily occur at modest temperatures (e.g. 250°C) over catalysts of mild activity such as copper supported on zinc oxide. Notice also that carbon monoxide does not feature as a principal product of methanol reforming. This makes methanol reformate particularly suited to PEM fuel cells, where carbon monoxide, even at the ppm level, can cause substantial losses in performance due to poisoning of the platinum catalyst. However, it is important to note that although carbon monoxide does not feature in reaction (5.4), this does not mean that it is not produced at all. The water gas shift reaction of (5.3) is reversible, and carbon monoxide is produced in small quantities. The result is that the carbon monoxide removal methods described below are still needed with a methanol reformer used with a PEM fuel cell.

5.2.3 Partial oxidation and autothermal reforming

As an alternative to steam reforming, methane and other hydrocarbons may be converted to hydrogen for fuel cells via partial oxidation (POX):

$$CH_4 + \tfrac{1}{2}O_2 \longrightarrow CO + 2H_2 \qquad [\Delta H = -247\,\text{kJ mol}^{-1}] \qquad (5.5)$$

$$C_8H_{18} + 4O_2 \longrightarrow 8CO + 9H_2 \qquad (5.6)$$

Partial oxidation can be carried out at high temperatures (typically 1200 to 1500°C) without a catalyst, but this is not practical in small mobile systems. If the temperature is reduced, and a catalyst employed then the process becomes known as Catalytic Partial Oxidation (CPO). Catalysts for CPO tend to be supported platinum-metal or nickel based.

It should be noted that reactions (5.5) and (5.6) produce less hydrogen per molecule of fuel than reactions (5.1) or (5.2). This means that partial oxidation (either non-catalytic or catalysed) is less efficient than steam reforming for fuel cell applications. Another disadvantage of partial oxidation occurs when air is used to supply the oxygen. This results in a lowering of the partial pressure of hydrogen at the fuel cell, because of the presence of the nitrogen, which further dilutes the hydrogen fuel. This in turn results in a lowering of the cell voltage, again resulting in a lowering of system efficiency. To offset these negative aspects, a key advantage of partial oxidation is that it does not require steam.

Autothermal reforming is another commonly used term in fuel processing. This usually describes a process in which both steam and oxidant (oxygen, or more normally air) are fed with the fuel to a catalytic reactor. It can therefore be considered as a combination of POX and the steam reforming processes already described. The basic idea of autothermal reforming is that both the *endothermic* steam reforming reaction (5.1) or (5.2) and the *exothermic* POX reaction of (5.5) or (5.6) occur together, so that no heat needs to be supplied or removed from the system. However, there is some confusion in the literature between the terms *partial oxidation* and *autothermal reforming*. Joensen and Rostrup-Nielsen (2002) have published a review which explains the issues in some detail.

The advantages of autothermal reforming and CPO are that less steam is needed than with conventional reforming and that all of the heat for the reforming reaction is provided by partial combustion of the fuel. This means that no complex heat management engineering is required, resulting in a simple system design. This is particularly attractive for mobile applications.

5.2.4 Further fuel processing: carbon monoxide removal

A steam reformer reactor running on natural gas and operating at atmospheric pressure with an outlet temperature of 800°C produces a gas comprising some 75% hydrogen, 15% carbon monoxide and 10% carbon monoxide on a dry basis. For the PEM fuel cell the carbon monoxide content must be reduced to much lower levels. Similarly, even the product from a methanol reformer operating at about 200°C will have at least 0.1% carbon monoxide content, depending on pressure and water content. The problem of reducing the carbon monoxide content of reformed gas streams is thus very important.

We have seen that the water gas shift reaction:

$$CO + H_2O \longleftrightarrow CO_2 + H_2 \qquad (5.7)$$

takes place at the same time as the basic steam reforming reaction. However, the thermodynamics of the reaction are such that higher temperatures favour the production of carbon monoxide, and shift the equilibrium to the left. The first approach is thus to *cool* the product gas from the steam reformer and pass it through a reactor containing catalyst, which promotes the shift reaction. This has the effect of converting carbon monoxide into carbon dioxide. Depending on the reformate composition more than one shift reactor may be needed, and two reactors is the norm. Such systems will give a carbon monoxide concentration of about 2500–5000 ppm, which exceeds the limit for PEM fuel cells by a factor of about 100. It is similar to the CO content in the product from a methanol reformer.

For PEM fuel cells, further carbon monoxide removal is essential after the shift reactors. This is usually done in one of four ways.

In the **selective oxidation reactor** a small amount of air (typically around 2%) is added to the fuel stream, which then passes over a precious metal catalyst. This catalyst preferentially absorbs the carbon monoxide, rather than the hydrogen, where it reacts with the oxygen in the air. As well as the obvious problem of cost, these units need to be very carefully controlled. There is the presence of hydrogen, carbon monoxide and oxygen, at an elevated temperature, with a noble metal catalyst. Measures must be taken to ensure that an explosive mixture is not produced. This is a special problem in cases where the flowrate of the gas is highly variable, such as with a PEMFC on a vehicle.

The **methanation** of the carbon monoxide is an approach that reduces the danger of producing explosive gas mixtures. The reaction is the opposite of the steam reformation reaction of equation (5.1):

$$CO + 3H_2 \longrightarrow CH_4 + H_2O \qquad (\Delta H = -206 \, \text{kJ.mol}^{-1})$$

This method has the obvious disadvantage that hydrogen is being consumed, and so the efficiency is reduced. However, the quantities involved are small; we are reducing the carbon monoxide content from about 0.25%. The methane does not poison the fuel cell, but simply acts as a diluent. Catalysts are available which will promote this reaction so that at about 200°C the carbon monoxide levels will be less than 10 ppm. The catalysts will also ensure that any unconverted methanol is reacted to methane, hydrogen or carbon dioxide.

Palladium/platinum membranes can be used to separate and purify the hydrogen. This is a mature technology that has been used for many years to produce hydrogen of exceptional purity. However, these devices are expensive.

Pressure swing absorption (PSA): in this process, the reformer product gas passed into a reactor containing absorbent material. Hydrogen gas is preferentially absorbed on this material. After a set time the reactor is isolated and the feed gas is diverted into a parallel reactor. At this stage the first reactor is depressurised, allowing pure hydrogen to desorb from the material. The process is repeated and the two reactors are alternately pressurised and depressurised. This process can be made to work well, but adds considerably to the bulk, cost and control problems of the system.

Currently none of these systems has established itself as the preferred option. They have the common feature that they add considerably to the cost and complexity of the fuel processing systems.

5.2.5 Practical fuel processing for mobile applications

The special features of onboard fuel processors for mobile applications are that they need:

- to be compact (both in weight and volume);
- to be capable of starting up quickly;
- to be able to follow demand rapidly and operate efficiently over a wide operating range;
- to be capable of delivering low-CO content gas to the PEM stack;
- to emit very low levels of pollutants.

Over the past few years, research and development of fuel processing for mobile applications, as well as small scale stationary applications, has mushroomed. Many organisations are developing proprietary technology, but almost all of them are based on the options outlined above, namely steam reforming, CPO, or autothermal reforming.

Companies such as Arthur D. Little have been developing reformers aimed at utilising gasoline type hydrocarbons (Teagan *et al.* 1998). The company felt that the adoption of gasoline as a fuel for FCVs would be likely to find favour amongst oil companies, since the present distribution systems can be used. Indeed Shell have demonstrated their own CPO technology on gasoline and ExxonMobil in collaboration with GM have also been developing a gasoline fuel processor. Arthur D. Little spun out its reformer development into Epyx which later teamed up with the Italian company De Nora, to form the fuel cell company Nuvera. In the Nuvera fuel processing system the required heat of reaction for the reforming is provided by in situ oxidising a fraction of the feedstock in a combustion (POX) zone. A nickel-based catalyst bed following the POX zone is the key to achieving full fuel conversion for high efficiency. The POX section operates at relatively high temperatures (1100–1500°C) whereas the catalytic reforming operates in the temperature range 800–1000°C. The separation of the POX and catalytic zones allows a relatively pure gas to enter the reformer, permitting the system to accommodate a variety of fuels. Shift reactors (high and low temperature) convert the product gas from the reformer so that the exit concentration of CO is less than 1%. As described earlier, an additional CO-removal stage is therefore needed to achieve the CO levels necessary for a PEM fuel cell. When designed for gasoline, the fuel processor also includes a compact desulphurisation bed integrated within the reactor vessel prior to the low temperature shift.

Johnson Matthey have demonstrated their HotSpot reactor on reformulated gasoline (Ellis *et al.* 2001). They built a 10 kW fuel processor which met their technical targets, but they also addressed issues relating to mass manufacture; their work has identified areas that will require further work to enable gasoline reforming to become a commercial reality. These included:

- hydrogen storage for start-up and transients;
- an intrinsically safe afterburner design with internal temperature control and heat exchange that can cope with transients;
- effect of additives on fuels;
- better understanding of the issues relating to sulphur removal from fuels at source;
- improved sulphur trapping and regeneration strategies.

Johnson Matthey are now engaged in a commercialisation programme for their technology. The pace of development is now such that in April 2001, GM demonstrated their own gasoline fuel processor in a Chevrolet 2–10 pickup truck, billed as the world's first gasoline-fed fuel cell electric vehicle. With the rapid developments being made in this area it remains to be seen which of the various fuel processing systems will become economically viable in the future.

One way to side-step all of the problems associated with onboard fuel processing is to make the fuel processing plant stationary, and to store the hydrogen produced, which can be loaded onto the mobile system as required. In fact, this is may well be the preferred option for some applications, such as buses. However, as ever, solving one problem creates others, and the problems of storing hydrogen are quite severe. These are dealt with in Sections 5.3 and 5.4 below.

5.3 Hydrogen Storage I: Storage as Hydrogen

5.3.1 Introduction to the problem

The difficulties arise because although hydrogen has one of the highest specific energies (energy per kilogram), which is why it is the fuel of choice for space missions, its density is very low, and it has one of the lowest energy densities (energy per cubic metre). This means that to get a large mass of hydrogen into a small space very high pressures have to be used. A further problem is that, unlike other gaseous energy carriers, it is very difficult to liquefy. It cannot be simply compressed, in the way that LPG or butane can. It has to be cooled down to about 22 K, and even in liquid form its density is really very low, 71 kg.m^{-3}.

Although hydrogen can be stored as a compressed gas or a liquid, there are other methods that are being developed. Chemical methods can also be used. These are considered in the next section. The methods of storing hydrogen that will be described in this section are: compression in gas cylinders, storage as a cryogenic liquid, storage in a metal absorber as a reversible metal hydride, and storage in carbon nanofibres.

None of these methods is without considerable problems, and in each situation their advantages and disadvantages will work differently. However, before considering them in detail we must address the vitally important issue of safety in connection with storing and using hydrogen.

5.3.2 Safety

Hydrogen is a unique gaseous element, possessing the lowest molecular weight of any gas. It has the highest thermal conductivity, velocity of sound, mean molecular velocity, and the lowest viscosity and density of all gases. Such properties lead hydrogen to have a leak rate through small orifices faster than all other gases. Hydrogen leaks 2.8 times faster than methane and 3.3 times faster than air. In addition hydrogen is a highly volatile and flammable gas, and in certain circumstances hydrogen and air mixtures can detonate. The implications for the design of fuel cell systems are obvious, and safety considerations must feature strongly.

Table 5.1 gives the key properties relevant to safety of hydrogen and two other gaseous fuels widely used in homes, leisure and business: methane and propane. From this table the major problem with hydrogen appears to be the minimum ignition energy, apparently indicating that a fire could be started very easily. However, all these energies are in fact very low, lower than those encountered in most practical cases. A spark can ignite any of these fuels. Furthermore, against this must be set the much higher minimum concentration needed for detonation, 18% by volume. The lower concentration limit for ignition is much the same as for methane, and a considerably lower concentration of propane is needed. The ignition temperature for hydrogen is also noticeably higher than for the other two fuels.

Hydrogen therefore needs to be handled with care. Systems need to be designed with the lowest possible chance of any leaks, and should be monitored for such leaks regularly. However, it should be made clear that, all things considered, hydrogen is no more dangerous, and in some respects it is rather less dangerous than other commonly used fuels.

5.3.3 The storage of hydrogen as a compressed gas

Storing hydrogen gas in pressurised cylinders is the most technically straightforward method, and the most widely used for small amounts of the gas. Hydrogen is stored in this way at thousands of industrial, research and teaching establishments, and in most locations local companies can readily supply such cylinders in a wide range of sizes. However, in these applications the hydrogen is nearly always a chemical reagent in some analytical or production process. When we consider using and storing hydrogen in this way as an energy vector, then the situation appears less satisfactory.

Table 5.1 Properties relevant to safety for hydrogen and two other commonly used gaseous fuels

	Hydrogen	Methane	Propane
Density, $kg.m^{-3}$ at NTP	0.084	0.65	2.01
Ignition limits in air, volume % at NTP	4.0 to 77	4.4 to 16.5	1.7 to 10.9
Ignition temperature, °C	560	540	487
Min. ignition energy in air, MJ	0.02	0.3	0.26
Max. combustion rate in air, ms^{-1}	3.46	0.43	0.47
Detonation limits in air, volume %	18 to 59	6.3 to 14	1.1 to 1.3
Stoichiometric ratio in air	29.5	9.5	4.0

Two systems of pressurised storage are compared in Table 5.2. The first is a standard steel alloy cylinder at 200 bar, of the type commonly seen in laboratories. The second is for larger scale hydrogen storage on a bus, as described by Zieger (1994). This tank is constructed with a 6 mm thick aluminium inner liner, around which is wrapped a composite of aramide fibre and epoxy resin. This material has a high ductility, which gives it good burst behaviour, in that it rips apart rather than disintegrating into many pieces. The burst pressure is 1200 bar, though the maximum pressure used is 300 bar.[2]

The larger scale storage system is, as expected, a great deal more efficient. However, this is slightly misleading. These large tanks have to be held in the vehicle, and the weight needed to do this should be taken into account. In the bus described by Zieger (1994), which used hydrogen to drive an internal combustion engine, 13 of these tanks were mounted in the roof space. The total mass of the tanks and the bus structure reinforcements is 2550 kg, or 196 kg per tank. This brings down the 'storage efficiency' of the system to 1.6%, not so very different from the steel cylinder. Another point is that in both systems we have ignored the weight of the connecting valves, and of any pressure-reducing regulators. For the 2 L steel cylinder system this would typically add about 2.15 kg to the mass of the system, and reduce the storage efficiency to 0.7% (Kahrom 1999).

The reason for the low mass of hydrogen stored, even at such very high pressures, is of course its low density. The density of hydrogen gas at normal temperature and pressure is $0.084 \, \text{kg.m}^{-3}$, compared to air, which has about $1.2 \, \text{kg.m}^{-3}$. *Usually less than 2% of the storage system mass is actually hydrogen itself.*

The metal that the pressure vessel is made from needs very careful selection. Hydrogen is a very small molecule, of high velocity, and so it is capable of diffusing into materials that are impermeable to other gases. This is compounded by the fact that a very small fraction of the hydrogen gas molecules may dissociate on the surface of the material. Diffusion of atomic hydrogen into the material may then occur which can affect the mechanical performance of materials in many ways. Gaseous hydrogen can build up in internal blisters in the material, which can lead to crack promotion (hydrogen-induced cracking). In carbonaceous metals such as steel the hydrogen can react with carbon, forming entrapped CH_4 bubbles. The gas pressure in the internal voids can generate an internal stress high enough to fissure, crack or blister the steel. The phenomenon is well

Table 5.2 Comparative data for two cylinders used to store hydrogen at high pressure. The first is a conventional steel cylinder, the second a larger composite tank for use on a hydrogen powered bus

	2 L steel, 200 bar	147 L composite, 300 bar
Mass of empty cylinder	3.0 kg	100 kg
Mass of hydrogen stored	0.036 kg	3.1 kg
Storage efficiency (% mass H_2)	1.2%	3.1%
Specific energy	$0.47 \, \text{kWh.kg}^{-1}$	$1.2 \, \text{kWh.kg}^{-1}$
Volume of tank (approx.)	2.2 L (0.0022 m^3)	220 L (0.22 m^3)
Mass of H_2 per litre	$0.016 \, \text{kg.L}^{-1}$	$0.014 \, \text{kg.L}^{-1}$

---

[2] It should be noted that at present composite cylinders have about three times the cost of steel cylinders of the same capacity.

known and is termed hydrogen embrittlement. Certain chromium-rich steels and Cr-Mo alloys have been found that are resistant to hydrogen embrittlement. Composite reinforced plastic materials are also used for larger tanks, as has been outlined above.

As well as the problem of very high mass, there are considerable safety problems associated with storing hydrogen at high pressure. A leak from such a cylinder would generate very large forces as the gas is propelled out. It is possible for such cylinders to become essentially jet-propelled torpedoes, and to inflict considerable damage. Furthermore, vessel fracture would most likely be accompanied by autoignition of the released hydrogen and air mixture, with an ensuing fire lasting until the contents of the ruptured or accidentally opened vessel are consumed (Hord 1978). Nevertheless, this method is widely and safely used, provided that the safety problems, especially those associated with the high pressure, are avoided by correctly following the due procedures. In vehicles, for example, pressure relief valves or rupture discs are fitted which will safely vent gas in the event of a fire for example. Similarly, pressure regulators attached to hydrogen cylinders are fitted with flame-traps to prevent ignition of the hydrogen.

The main advantages of storing hydrogen as a compressed gas are: simplicity, indefinite storage time, and no purity limits on the hydrogen. Designs for very high-pressure cylinders can be incorporated into vehicles of all types. In the fuel cell bus of Figures 1.16 and 11.6 they are in the roof. Figure 5.4 shows the design of a modern very high-pressure hydrogen storage system by General Motors, and its location in the fuel cell powered vehicle can be seen in the picture in the background.

5.3.4 Storage of hydrogen as a liquid

The storage of hydrogen as a liquid (commonly called LH_2), at about 22 K, is currently the only widely used method of storing large quantities of hydrogen. A gas cooled to the liquid

Figure 5.4 General Motors very high pressure hydrogen gas cylinder

state in this way is known as a cryogenic liquid. Large quantities of cryogenic hydrogen are currently used in processes such as petroleum refining and ammonia production. Another notable user is NASA, which has huge $3200 \, m^3$ (850 000 US gallon) tanks to ensure a continuous supply for the space programme.

The hydrogen container is a large, strongly reinforced vacuum (or Dewar) flask. The liquid hydrogen will slowly evaporate, and the pressure in the container is usually maintained below 3 bar, though some larger tanks may use higher pressures. If the rate of evaporation exceeds the demand, then the tank is occasionally vented to make sure the pressure does not rise too high. A spring loaded valve will release, and close again when the pressure falls. The small amounts of hydrogen involved are usually released to the atmosphere, though in very large systems it may be vented out through a flare stack and burnt. As a back-up safety feature a rupture disc is usually also fitted. This consists of a ring covered with a membrane of controlled thickness, so that it will withstand a certain pressure. When a safety limit is reached, the membrane bursts, releasing the gas. However, the gas will continue to be released until the disc is replaced. This will not be done until all the gas is released, and the fault rectified.

When the LH_2 tank is being filled, and when fuel is being withdrawn, it is most important that air is not allowed into the system, otherwise an explosive mixture could form. The tank should be purged with nitrogen before filling.

Although usually used to store large quantities of hydrogen, considerable work has gone into the design and development of LH_2 tanks for cars, though this has not been directly connected with fuel cells. BMW, among other automobile companies, has invested heavily in hydrogen powered internal combustion engines, and these have used LH_2 as the fuel. Such tanks have been through very thorough safety trials. The tank used in their hydrogen powered cars is cylindrical in shape, and is of the normal double wall, vacuum or Dewar flask type of construction. The walls are about 3 cm thick, and consist of 70 layers of aluminium foil interlaced with fibre-glass matting. The maximum operating pressure is 5 bar. The tank stores 120 litres of cryogenic hydrogen. The density of LH_2 is very low, about $71 \, kg.m^{-3}$, so 120 litres is only 8.5 kg (Reister and Strobl 1992). The key figures are shown in Table 5.3.

The hydrogen fuel feed systems used for car engines cannot normally be applied unaltered to fuel cells. One notable difference is that in LH_2 powered engines the hydrogen is often fed to the engine still in the liquid state. If it is a gas, then being at a low temperature is an advantage, as it allows a greater mass of fuel/air mixture into the engine. For fuel cells, the hydrogen will obviously need to be a gas, and pre-heated as

Table 5.3 Details of a cryogenic hydrogen container suitable for cars

Mass of empty container	51.5 kg
Mass of hydrogen stored	8.5 kg
Storage efficiency (% mass H_2)	14.2%
Specific energy	5.57 $kWh.kg^{-1}$
Volume of tank (approx.)	0.2 m^3
Mass of H_2 per litre	0.0425 $kg.L^{-1}$

well. However, this is not a very difficult technical problem, as there is plenty of scope for using waste heat from the cell via heat exchangers.

One of the problems associated with cryogenic hydrogen is that the liquefaction process is very energy-intensive. Several stages are involved. The gas is firstly compressed, and then cooled to about 78 K using liquid nitrogen. The high pressure is then used to further cool the hydrogen by expanding it through a turbine. An additional process is needed to convert the H_2 from the isomer where the nuclear spins of both atoms are parallel (ortho-hydrogen) to that where they are anti-parallel (para-hydrogen). This process is exothermic, and if allowed to take place naturally would cause boil-off of the liquid. According to figures provided by a major hydrogen producer, and given by Eliasson and Bossel (2002), the energy required to liquefy the gas under the *very best of circumstances* is about 25% of the specific enthalpy or heating value of the hydrogen. This is for modern plants liquefying over 1000 kilograms per hour. For plants working at about 100 kg.h^{-1}, hardly a small rate, the proportion of the energy lost rises to about 45%. In overall terms then, this method is a highly inefficient way of storing and transporting energy.

In addition to the regular safety problems with hydrogen, there are a number of specific difficulties concerned with cryogenic hydrogen. Frostbite is a hazard of concern. Human skin can easily become frozen or torn if it comes into contact with cryogenic surfaces. All pipes containing the fluid must be insulated, as must any parts in good thermal contact with these pipes. Insulation is also necessary to prevent the surrounding air from condensing on the pipes, as an explosion hazard can develop if liquid air drips onto nearby combustibles. Asphalt, for example, can ignite in the presence of liquid air. (Concrete paving is used around static installations.) Generally though, the hazards of hydrogen are somewhat less with LH_2 than with pressurised gas. One reason is that if there is a failure of the container, the fuel tends to remain in place, and vent to the atmosphere more slowly. Certainly, LH_2 tanks have been approved for use in cars in Europe.

5.3.5 Reversible metal hydride hydrogen stores

The reader might well question the inclusion of this method in this section, rather than with the chemical methods that follow. However, although the method is chemical in its operation, that is not in any way apparent to the user. No reformers or reactors are needed to make the systems work. They work exactly like a hydrogen 'sponge' or 'absorber'. For this reason it is included here.

Certain metals, particularly mixtures (alloys) of titanium, iron, manganese, nickel, chromium, and others, can react with hydrogen to form a metal hydride in a very easily controlled reversible reaction. The general equation is:

$$M + H_2 \longleftrightarrow MH_2 \qquad (5.8)$$

To the right, the reaction of (5.8) is mildly exothermic. To release the hydrogen, then, small amounts of heat must be supplied. However, metal alloys can be chosen for the hydrides so that the reaction can take place over a wide range of temperatures and pressures. In particular, it is possible to choose alloys suitable for operating at around atmospheric pressure, and room temperature.

The system works as follows. Hydrogen is supplied at a little above atmospheric pressure to the metal alloy, inside a container. The reaction of (5.8) proceeds to the right, and the metal hydride is formed. This is mildly exothermic, and in large systems some cooling will need to be supplied, but normal air cooling is often sufficient. This stage will take a few minutes, depending on the size of the system, and if the container is cooled. It will take place at approximately constant pressure.

Once all the metal has reacted with the hydrogen, then the pressure will begin to rise. This is the sign to disconnect the hydrogen supply. The vessel, now containing the metal hydride, will then be sealed. Note that the hydrogen is only stored at modest pressure, typically up to 5 bar.

When the hydrogen is needed, the vessel is connected to, for example, the fuel cell. The reaction of (5.8) then proceeds to the left, and hydrogen is released. If the pressure rises above atmospheric, the reaction will slow down or stop. The reaction is now endothermic, so energy must be supplied. This is supplied by the surroundings; the vessel will cool slightly as the hydrogen is given off. It can be warmed slightly to increase the rate of supply, using, for example, warm water or the air from the fuel cell cooling system.

Once the reaction has completed, and all the hydrogen has been released, then the whole procedure can be repeated. *Note that we have already met this process*, when we looked at the metal hydride battery in Chapter 2; the same process is used to store hydrogen directly on the negative electrode.

Usually several hundred charge/discharge cycles can be completed. However, rather like rechargeable batteries, these systems can be abused. For example, if the system is filled at high pressure, the charging reaction will proceed too fast, and the material will get too hot, and will be damaged. Another important problem is that the containers are damaged by impurities in the hydrogen; the metal absorbers will react permanently with them. So a high purity hydrogen, at least 99.999% pure, must be used.

Although the hydrogen is not stored at pressure, the container must be able to withstand a reasonably high pressure, as it is likely to be filled from a high pressure supply, and allowance must be made for human error. For example, the unit shown in Figure 5.5 will be fully charged at a pressure of 3 bar, but the container can withstand 30 bar. The container will also need valves and connectors. Even taking all these into account impressive practical devices can be built. In Table 5.4 gives details of the small 20 SL holder for applications such as portable electronics equipment, manufactured by GfE Metalle und Materialien GMBH of Germany, and shown in Figure 5.5. The volumetric measure, mass of hydrogen per litre, is nearly as good as for LH_2, and the gravimetric measure is not a great deal worse than for compressed gas, and very much the same as for a small compressed cylinder. Larger systems have very similar performance.

One of the main advantages of this method is its safety. The hydrogen is not stored at a significant pressure, and so cannot rapidly and dangerously discharge. Indeed, if the valve is damaged, or there is a leak in the system, the temperature of the container will fall, which will inhibit the release of the gas. The low pressure greatly simplifies the design of the fuel supply system. It thus has great promise for a very wide range of applications where small quantities of hydrogen are stored. It is also particularly suited to applications where weight is not a problem, but space is.

Table 5.4 Details of a small metal hydride hydrogen container suitable for portable electronics equipment

Mass of empty container	0.26 kg
Mass of hydrogen stored	0.0017 kg
Storage efficiency (% mass H_2)	0.65%
Specific energy	0.26 kWh.kg^{-1}
Volume of tank (approx.)	0.061
Mass of H_2 per litre	0.028 kg.L^{-1}

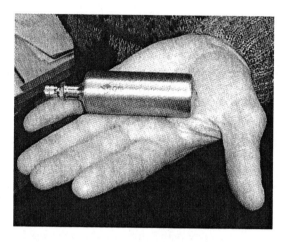

Figure 5.5 Metal hydride stores can be made quite small, as this example shows

The disadvantages are particularly noticeable where larger quantities of hydrogen are to be stored, for example in vehicles! The specific energy is poor. Also, the problem of the heating during filling and cooling during release of hydrogen becomes more acute. Large systems have been tried for vehicles, and a typical refill time is about one hour for an approximately 5 kg tank. The other major disadvantage is that usually very high purity hydrogen must be used, otherwise the metals become contaminated, as they react irreversibly with the impurities.

5.3.6 Carbon nanofibres

In 1998 a paper was published on the absorption of hydrogen in carbon nanofibres (Chambers *et al.* 1998). The authors presented results suggesting that these materials could absorb in excess of 67% hydrogen by weight, a storage capacity far in excess of any of the others we have described so far. This set many other workers on the same trail. However, it would be fair to say that no-one has been able to repeat this type of performance, and methods by which errors could be made in the measurements have been suggested. Nevertheless, other workers have shown fairly impressive storage capability with carbon nanofibres, and this is certainly one to watch for the future (see Chapter 8 of Larminie and Dicks (2003)).

5.3.7 Storage methods compared

Table 5.5 shows the range of gravimetric and volumetric hydrogen storage measures for the three systems described above that are available now. Obviously these figures cannot be used in isolation; they don't include cost, for example. Safety aspects do not appear in this table either. The cryogenic storage method has the best figures.

Table 5.5 Data for comparing methods of storing hydrogen fuel

Method	Gravimetric storage efficiency, % mass hydrogen	Volumetric mass (in kg) of hydrogen per litre
Pressurised gas	0.7–3.0	0.015
Reversible metal hydride	0.65	0.028
Cryogenic liquid	14.2	0.040

5.4 Hydrogen Storage II: Chemical Methods

5.4.1 Introduction

None of the methods for storing hydrogen outlined in Section 5.3 is entirely satisfactory. Other approaches that are being developed rely on the use of chemical 'hydrogen carriers'. These could also be described as 'man-made fuels'. There are many compounds that can be manufactured to hold, for their mass, quite large quantities of hydrogen. To be useful these compounds must pass three tests:

1. It must be possible to very easily make these compounds give up their hydrogen, otherwise there is no advantage over using a reformed fuel in one of the ways already outlined in Section 5.2.
2. The manufacturing process must be simple and use little energy; in other words the energy and financial costs of putting the hydrogen into the compound must be low.
3. They must be safe to handle.

A large number of chemicals that show promise have been suggested or tried. Some of these, together with their key properties, are listed in Table 5.6. Some of them do not warrant a great deal of consideration, as they easily fail one or more of the three tests above. Hydrazine is a good example. It passes the first test very well, and it has been used in demonstration fuel cells with some success. However, hydrazine is both highly toxic and very energy-intensive to manufacture, and so fails the second and third tests.

Table 5.6 Liquids that might be used to locally store hydrogen gas for fuel cells

Name	Formula	Percent H_2	Density, $kg.L^{-1}$	Vol. (l) to store 1 kg H_2	Notes
Liquid H_2	H_2	100	0.07	14	Cold, $-252°C$
Ammonia	NH_3	17.8	0.67	8.5	Toxic, 100 ppm
Liquid methane	CH_4	25.1	0.415	9.6	Cold, $-175°C$
Methanol	CH_3OH	12.5	0.79	10	
Ethanol	C_2H_5OH	13.0	0.79	9.7	
Hydrazine	N_2H_4	12.6	1.01	7.8	Highly toxic
30% sodium borohydride solution	$NaBH_4 + H_2O$	6.3	1.06	15	Expensive, but works well

Nevertheless, several of the compounds of Table 5.6 are being considered for practical applications, and will be described in more detail here.

5.4.2 Methanol

Methanol is the 'man-made' carrier of hydrogen that is attracting the most interest among fuel cell developers. As we saw in Section 5.2, methanol can be reformed to hydrogen by steam reforming, according to the following reaction:

$$CH_3OH + H_2O \longrightarrow CO_2 + 3H_2 \qquad (5.9)$$

The equipment is much more straightforward, though the process is not so efficient, if the partial oxidation route is used, for which the reaction is:

$$2CH_3OH + O_2 \longrightarrow 2CO_2 + 4H_2 \qquad (5.10)$$

The former would yield 0.188 kg of hydrogen for each kg of methanol, the latter 0.125 kg of hydrogen for each kg of methanol. We have also seen in Section 5.2 that *autothermal* reformers use a combination of both these reactions, and this attractive alternative would provide a yield somewhere between these two figures. The key point is that whatever reformation reaction is used (equation (5.9) or (5.10) the reaction takes place at temperatures around 250°C, which is far less than those needed for the reformation of gasoline, as described in Section 5.2 (equations (5.2) or (5.6)). Also, the amount of carbon monoxide produced is far less, which means that far less chemical processing is needed to remove it. All that is needed is one of the four carbon monoxide clean-up systems outlined in Section 5.2.4.

Leading developers of methanol reforming for vehicles at present are Excellsis Fuel cell Engines (DaimlerChrysler), General Motors, Honda, International Fuel Cells, Mitsubishi, Nissan, Toyota, and Johnson Matthey. Most are using steam reforming although some organisations are also working on partial oxidation. DaimlerChrysler developed a methanol processor for the NeCar 3 experimental vehicle. This was demonstrated in September

Table 5.7 Characteristics of the methanol processor for NeCar 3 (Kalhammer et al. 1998)

Maximum unit size	50 kWe
Power density	1.1 kW$_e$.L^{-1} (reformer = 20 L, combustor = 5 L, CO selective oxidiser 20 L)
Specific power	0.44 kW$_e$.kg^{-1} (reformer = 34 kg, combustor = 20 kg, CO sel. oxidiser = 40 kg)
Energy efficiency	not determined
Methanol conversion Efficiency	98–100%
Turn-down ratio	20 to 1
Transient response	<2 s

1997 as the world's first methanol-fuelled fuel cell car. It was used in conjunction with a Ballard 50 kW fuel cell stack. Characteristics of the methanol processor are given in Table 5.7.

Since the NeCar3 demonstration, DaimlerChrysler and Excellsis have been working with BASF to develop a more advanced catalytic reformer system for their vehicles. In November 2000 DaimlerChrysler launched the NeCar5 which, together with a Jeep Commander vehicle, represents state-of-the-art in methanol fuel cell vehicles. In the past six years, the fuel cell drive system has been shrunk to such an extent that it presently requires no more space than a conventional drive system. The Necar 5 therefore has the full complement of seats and interior space as a conventional gasoline-fuelled internal combustion engine car. The car is based on the A-Class Mercedes design and the methanol reformer is under the passenger compartment, as illustrated in Figure 5.6. The NeCar 5 uses a Ballard 75 kW Mk 9 stack, giving an impressive top speed of over 150 km/h.

However, whatever reformer is used, full utilisation is not possible; it never is with gas mixtures containing carbon dioxide, as there must still be some hydrogen in the exit gas, as explained in Section 5.2.1. Also, in the case of steam reforming, some of the product hydrogen is needed to provide energy for the reforming reaction. If we assume that the hydrogen utilisation can be 75%, then we can obtain 0.14 kg of hydrogen for each kg of methanol. We can speculate that a 40 litre tank of methanol might be used, with a reformer of about the same size and weight as the tank. Such a system should be possible in the reasonably near term, and would give the figures of Table 5.8.

The potential figures show why methanol systems are looked on with such favour, and why they are receiving a great deal of attention for systems of power above about 10 W right through to tens of kilowatts.

We should note that ethanol, according to the figures of Table 5.6, should be just as promising as methanol as a hydrogen carrier. Its main disadvantage is that the equivalent reformation reactions of equations (5.9) and (5.10) do not proceed nearly so readily, making the reformer markedly larger, more expensive, less efficient and more difficult to control. Ethanol is also usually somewhat more expensive. All these disadvantages more than counter its very slightly higher hydrogen content.

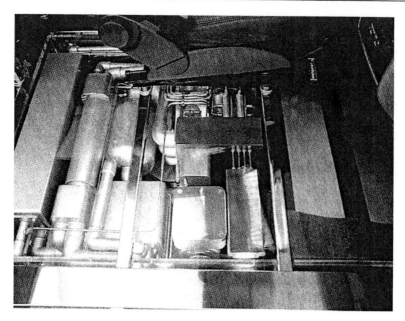

Figure 5.6 Packaging of the NeCar 5 methanol fuel processor

Table 5.8 Speculative data for a hydrogen source, storing 40 L (32 kg) of methanol

Mass of reformer and tank	64 kg
Mass of hydrogen stored[a]	4.4 kg
Storage efficiency (% mass H_2)	6.9%
Specific energy	5.5 kWh.kg^{-1}
Volume of tank + reformer	0.08 m^3
Mass of H_2 per litre	0.055 kg.L^{-1}

[a] Assuming 75% conversion of available H_2 to usable H_2.

5.4.3 Alkali metal hydrides

An alternative to the reversible metal hydrides (Section 5.3.5) are alkali metal hydrides which react with water to release hydrogen, and produce a metal hydroxide. Bossel (1999) has described a system using calcium hydride which reacts with water to produce calcium hydroxide and release hydrogen:

$$CaH_2 + 2H_2O \longrightarrow Ca(OH)_2 + 2H_2 \tag{5.11}$$

It could be said that the hydrogen is being released from the water by the hydride.

Another method that is used commercially, under the trade name Powerballs, is based on sodium hydride. These are supplied in the form of polyethylene coated spheres of about 3 cm diameter. They are stored underwater, and cut in half when required to produce hydrogen. An integral unit holds the water, product sodium hydroxide, and a microprocessor controlled cutting mechanism that operates to ensure a continuous supply of hydrogen. In this case the reaction is:

$$NaH + H_2O \longrightarrow NaOH + H_2 \qquad (5.12)$$

This is a very simple way of producing hydrogen, and its energy density and specific energy can be as good or better than the other methods we have considered so far. Sodium is an abundant element, and so sodium hydride is not expensive. The main problems with these methods are:

- the need to dispose of a corrosive and unpleasant mixture of hydroxide and water; in theory, this could be recycled to produce fresh hydride, but the logistics of this would be difficult;
- the fact that the hydroxide tends to attract and bind water molecules, which means that the volumes of water required tend to be considerably greater than equations (5.11) and (5.12) would imply;
- the energy required to manufacture and transport the hydride is greater than that released in the fuel cell.

A further point is that the method does not stand very good comparison with metal air batteries. If the user is prepared to use quantities of water, and is prepared to dispose of water/metal hydroxide mixtures, then systems such as the aluminium/air or magnesium/air battery are preferable. With a salt water electrolyte, an aluminium/air battery can operate at 0.8 V at quite a high current density, producing three electrons for each aluminium atom. The electrode system is much cheaper and simpler than a fuel cell.

Nevertheless, the method compares quite well with the other systems in several respects. The figures in Table 5.9 are calculated for a self-contained system capable of producing 1 kg of hydrogen, using the sodium hydride system. The equipment for containing the water and gas, and the cutting and control mechanism is assumed to weigh 5 kg. There is three times as much water as equation (5.12) would imply is needed.

The storage efficiency compares well with other systems. This method may well have some niche applications where the disposal of the hydroxide is not a problem, though these are liable to be limited.

Table 5.9 Figures for a self-contained system producing 1 kg of hydrogen using water and sodium hydride

Mass of container and all materials	45 kg
Mass of hydrogen stored	1.0 kg
Storage efficiency (% mass H_2)	2.2%
Specific energy	0.87 kWh.kg^{-1}
Volume of tank (approx.)	50 L
Mass of H_2 per litre	0.020 kg.L^{-1}

5.4.4 Sodium borohydride

A good deal of interest has recently been shown in the use of sodium tetrahydridoborate, or sodium borohydride as it is usually called, as a chemical hydrogen carrier. This reacts with water to form hydrogen according to the reaction:

$$NaBH_4 + 2H_2O \longrightarrow 4H_2 + NaBO_2 \quad (\Delta H = -218 \, kJ.mol^{-1}) \quad (5.13)$$

This reaction does not normally proceed spontaneously, and solutions of $NaBH_4$ in water are quite stable. Some form of catalyst is usually needed. The result is one of the great advantages of this system: it is highly controllable. Millennium Cell Corp. in the USA has been actively promoting this system and has built demonstration vehicles running on both fuel cells and internal combustion engines using hydrogen made in this way. Companies in Europe, notably NovArs GmbH in Germany (Koschany 2001), have also made smaller demonstrators. Notable features of equation (5.13) are:

- it is exothermic, at the rate of 54.5 kJ per mole of hydrogen;
- hydrogen is the only gas produced, it is not diluted with carbon dioxide;
- if the system is warm, then water vapour will be mixed with the hydrogen, which is highly desirable for PEM fuel cell systems.

Although rather overlooked in recent years, $NaBH_4$ has been known as a viable hydrogen generator since 1943. The compound was discovered by the Nobel laureate Herbert C. Brown, and the story is full of interest and charm, but is well told by Prof. Brown himself (Brown 1992). Suffice to say that shortly before the end of the 1939–1945 war plans were well advanced to bulk-manufacture the compound for use in hydrogen generators by the US Army Signals Corps, when peace rendered this unnecessary. However, in the following years many other uses of sodium borohydride, notably in the paper processing industries, were discovered, and it is produced at the rate of about 5000 tonnes per year,[3] mostly using Brown's method, by Morton International (merged with Rohm and Haas in 1999).

If mixed with a suitable catalyst, $NaBH_4$ can be used in solid form, and water added to make hydrogen. The disadvantage of this method is that the material to be transported is a flammable solid, which spontaneously gives off H_2 gas if it comes into contact with water. This is obviously a safety hazard. It is possible to purchase sodium borohydride mixed with 7% cobalt chloride for this purpose. However, this is not the most practical way to use the compound.

Current work centres on the use of solutions. This has several advantages. Firstly the hydrogen source becomes a single liquid, no separate water supply is needed. Secondly this liquid is not flammable, and only mildly corrosive, unlike the solid form. The hydrogen releasing reaction of equation (5.13) is made to happen by bringing the solution into contact with a suitable catalyst. Removing the catalyst stops the reaction. The gas generation is thus very easily controlled, a major advantage in fuel cell applications.

The maximum practical solution strength used is about 30%. Higher concentrations are possible, but take too long to prepare, and are subject to loss of solid at lower temperatures.

---

[3] *Kirk-Othmer Encyclopedia of Chemical Technology*, Wiley.

The solution is made alkaline by the addition of about 3% sodium hydroxide, otherwise the hydrogen evolution occurs spontaneously. The 30% solution is quite thick, and so weaker solutions are sometimes used, even though their effectiveness as a hydrogen carrier is worse. One litre of a 30% solution will give 67 g of hydrogen, which equates to about 800 NL. This is a very good volumetric storage efficiency.

Generators using these solutions can take several forms. The principle is that to generate hydrogen the solution is brought into contact with a suitable catalyst, and that generation ceases when the solution is removed from the catalyst. Suitable catalysts include platinum and ruthenium, but other less expensive materials are effective, including iron oxide. Fuel cell electrodes make very good reactors for this type of generator.

A practical sodium borohydride system is shown in Figure 5.7. The solution is pumped over the reactor, releasing hydrogen. The motor driving the pump is turned on and off by a simple controller that senses the pressure of the hydrogen, and which turns it on when more is required. The solution is forced through the reactor, and so fresh solution is continually brought in contact with the catalyst. The rate of production is simply controlled by the duty cycle of the pump. The reaction takes place at room temperature, and the whole system is extremely simple when compared to any of the other generators that have been outlined in this section.

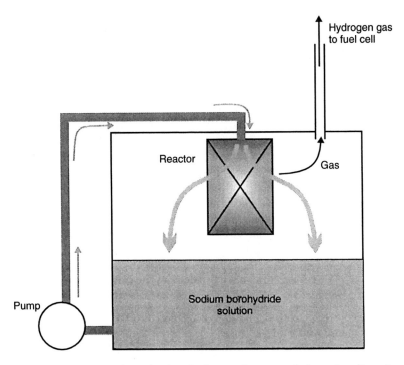

Figure 5.7 Example reactor for releasing hydrogen from a solution of sodium borohydride in water, stabilised with sodium hydroxide. The rate of production of hydrogen is controlled by varying the rate at which the solution is pumped over the reactor

However, when the solution is weak, the reaction rate will be much slower, and the system will behave differently. It is likely that it will not be practical to obtain highly efficient solution usage. Also, the solution cannot be renewed at the user's convenience; it must be completely replaced when the $NaBH_4$ has been used up, and at no other time.

Another method that is used is that of the Hydrogen on Demand® system promoted by Millennium Cell. This uses a single pass catalyst, rather than the re-circulation system of Figure 5.7. A major advantage of this is that the tank of fresh solution can be topped up at any time. A disadvantage is that two tanks are needed, the second one for the spent solution that has passed over the catalyst.

In terms of the general figure of merit, 'volume required to produce 1 kg of hydrogen', the 30% $NaBH_4$ solution is the worst of the liquid carriers shown in Table 5.6. However, it is competitive, and is only very slightly worse than pure liquid hydrogen. However, it has many advantages over the other technologies.

- it is arguably the safest of all the liquids to transport;
- apart from cryogenic hydrogen it is the only liquid that gives pure hydrogen as the product. This is very important, as it means this the only one where the product gas can be 100% utilised within the fuel cell;
- the reactor needed to release the hydrogen requires no energy, and can operate at ambient temperature and pressure;
- the rate of production of hydrogen can be simply controlled;
- the reactor needed to promote the hydrogen production reaction is very simple, far simpler than that needed for any of the other liquids;
- if desired, the product hydrogen gas can contain large quantities of water vapour, which is highly desirable for PEM fuel cells.

In order to compare a complete system, and produce comparative figures for gravimetric and volumetric storage efficiency, we need to speculate what a complete hydrogen generation system would be like. Systems have been built where the mass of the unit is about the same as the mass of the solution stored, and about twice the volume of solution held. So, a system that holds 1 litre of solution has a volume of about 2 litres, and weighs about 2 kg. Such a system would yield the figures shown in Table 5.10.

These figures are very competitive with all other systems. So what are the disadvantages? There are three main problems, the second two being related. The first is the problem of disposing of the borate solution. This is not unduly difficult, as it is not a

Table 5.10 Speculative data for a hydrogen source, storing 1.0 l of 30% $NaBH_4$, 3% NaOH and 67% H_2O solution

Mass of reformer, tank, and solution	2.0 kg
Mass of hydrogen stored	0.067 kg
Storage efficiency (% mass H_2)	3.35%
Specific energy	1.34 kWh.kg^{-1}
Volume of system (approx.)	2.0 L
Mass of H_2 per litre	0.036 kg.L^{-1}

Hydrogen Supply

hazardous substance. However, the other disadvantages are far more severe. The first is the cost. Sodium borohydride is an expensive compound. By simple calculation and reference to catalogues it can be shown that the cost of producing hydrogen this way is about $630 per kilogram.[4] This is over 100 times more expensive than using an electrolyser driven by grid-supplied electricity (Larminie 2002). At this sort of cost the system is not at all viable.

Linked to this problem of cost is the energy required to manufacture sodium borohydride. Using current methods this far exceeds the requirements of compounds such as methanol. Currently sodium borohydride is made from borax ($NaO.2B_2O_3.10H_2O$), a naturally occurring mineral with many uses that is mined in large quantities, and methanol. The aim is that the sodium metaborate produced by the hydrolysis reaction of equation (5.13) is recycled back to sodium borohydride. Table 5.11 shows the molar enthalpies of formation of the key compounds. It can be seen that such recycling will be a formidable challenge, requiring at least $788\,kJ.mol^{-1}$. However, the prize is *four* moles of hydrogen, so that is at least $197\,kJ.mol^{-1}$, which is not quite so daunting. Nevertheless, there are many problems to be overcome before such recycling is viable.

The companies, such as Millennium Cell Inc., who are hoping to commercialise this process are working hard on this problem of production cost, finance and energy. If they succeed there will be a useful hydrogen carrier, but until the costs come down by a factor of at least 10 then this method will only be suitable for special niche applications.

5.4.5 Ammonia

Ammonia is a colourless gas with a pungent choking smell that is easy to recognise. It is highly toxic. The molecular formula is NH_3, which immediately indicates its potential as a hydrogen carrier. It has many uses in the chemical industry, the most important being in the manufacture of fertiliser, which accounts for about 80% of the use of ammonia. It is also used in the manufacture of explosives. Ammonia is produced in huge quantities. The annual production is estimated at about 100 million tonnes, of which a little over 16 million is produced in the USA.[5]

Ammonia liquefies at $-33°C$, not an unduly low temperature, and can be kept in liquid form at normal temperature under its own vapour pressure of about 8 bar, not an unduly high pressure. Bulk ammonia is normally transported and stored in this form. However, it also readily dissolves in water – in fact it is the most water-soluble of all gases. The solution (ammonium hydroxide) is strongly alkaline, and is sometimes known as ammonia water or ammonia liquor. Some workers have built hydrogen generators using this as the

Table 5.11 Key thermodynamic data for sodium borohydride and borate

	$NaBH_4$	$NaBO_2$
Molar enthalpy of formation	$-189\,kJ.mol^{-1}$	$-977\,kJ.mol^{-1}$

---

[4] 2003 prices.
[5] Information provided by the Lousiana Ammonia Association, www.lammonia.com.

form of ammonia supplied, but this negates the main advantage of ammonia, which is its high hydrogen density, as well as adding complexity to the process.

Liquid ammonia is one of the most compact ways of storing hydrogen. In terms of volume needed to store 1 kg of hydrogen, it is better than almost all competing materials; see Table 5.6. Counter-intuitively, it is approximately 1.7 times as effective as liquid hydrogen. (This is because, even in liquid form, hydrogen molecules are very widely spaced, and LH_2 has a very low density.)

Table 5.6 shows ammonia to be the best liquid carrier, in terms of space to store 1 kg of hydrogen, apart from hydrazine, which is so toxic and carcinogenic that it is definitely not a candidate for regular use. However, the margin between the leading candidates is not very large. The figures ignore the large size of container that would be needed, especially in the case of ammonia, liquid methane and LH_2, though not for the key rival compound methanol.

Two other features of ammonia lie behind the interest in using it as a hydrogen carrier. The first is that large stockpiles are usually available, due to the seasonal nature of fertiliser use. The second is that ammonia prices are sometimes somewhat depressed due to an excess of supply over demand. However, when the details of the manufacture of ammonia, and its conversion back to hydrogen are considered, it becomes much less attractive.

Using ammonia as a hydrogen carrier involves the manufacture of the compound from natural gas and atmospheric nitrogen, the compression of the product gas into liquid form, and then, at the point of use, the dissociation of the ammonia back into nitrogen and hydrogen.

The production of ammonia involves the steam reformation of methane (natural gas), as outlined in Section 5.2. The reaction has to take place at high temperature, and the resulting hydrogen has to be compressed to very high pressure (typically 100 bar) to react with nitrogen in the Haber process. According to the Lousiana Ammonia Producers Association, who make about 40% of the ammonia produced in the USA, the efficiency of this process is about 60%. By this they mean that 60% of the gas used goes to provide hydrogen, and 40% is used to provide energy for the process. This must be considered a 'best case' figure, since there will no doubt be considerable use of electrical energy to drive pumps and compressors that is not considered here. The process is inherently very similar to methanol production; hydrogen is made from fuel, and is them reacted with another gas. In this case it is nitrogen instead of carbon dioxide. The process efficiencies and costs are probably similar.

The recovery of hydrogen from ammonia involves the simple dissociation reaction:

$$NH_3 \longrightarrow \tfrac{1}{2}N_2 + \tfrac{3}{2}H_2 \quad (\Delta H = +46.4 \text{ kJ.mol}^{-1}) \quad (5.14)$$

For this reaction to occur at a useful rate the ammonia has to be heated to between 600 and 800°C, and passed over a catalyst. Higher temperatures of about 900°C are needed if the output from the converter is to have remnant ammonia levels down to the ppm level. On the other hand the catalysts need not be expensive: iron, copper, cobalt and nickel are among many materials that work well. Systems doing this have been described in the literature (Kaye et al. 1997, Faleschini et al. 2000). The later paper has a good review of the catalysts that can be used.

The reaction is endothermic, as shown. However, this is not the only energy input required. The liquid ammonia absorbs large amounts of energy as it vaporises into a gas, which is why it is still quite extensively used as a refrigerant.

$$NH_3 \text{ (l)} \longrightarrow NH_3 \text{ (g)} \quad (\Delta H = +23.3 \text{ kJ.mol}^{-1}) \quad (5.15)$$

Once a gas at normal temperature, it then has to be heated, because the dissociation reaction only takes place satisfactorily at temperatures of around 800 to 900°C. For simplicity we will assume an 800°C temperature rise. The molar specific heat of ammonia is 36.4 J.mol^{-1}.kg^{-1}. So:

$$\Delta H = 800 \times 36.4 = 29.1 \text{ kJ.mol}^{-1}$$

This process results in the production of 1.5 moles of hydrogen, for which the molar enthalpy of formation (HHV) is -285.84 kJ.mol^{-1}. The best possible efficiency of this stage of the process is thus:

$$\frac{(285.84 \times 1.5) - (23.3 + 29.1 + 46.4)}{285.84 \times 1.5} = 0.77 = 77\%$$

This should be considered an upper limit of efficiency, as we have not considered the fact that the reformation process will involve heat losses to the surroundings. However, systems should be able to get quite close to this figure, since there is scope for using heat recovery, as the product gases would need to be cooled to about 80°C before entering the fuel cell. The vaporisation might also take place below ambient temperature, allowing some heat to be taken from the surroundings.

The corrosive nature of ammonia and ammonium hydroxide is another major problem. Water is bound to be present in a fuel cell. Any traces of ammonia left in the hydrogen and nitrogen product gas stream will dissolve in this water, and thus form an alkali (ammonium hydroxide) inside the cell. In small quantities, in an alkaline electrolyte fuel cell, this is tolerable. However, in the PEM fuel it would be fatal. This point is admitted by some proponents of ammonia, and is used by them as an advantage for alkaline fuel cells (Kordesch et al. 1999). Hydrogen from other hydrogen carriers such as methanol and methane also contains poisons, notably carbon monoxide. However, these can be removed, and do not permanently harm the cell, they just temporarily degrade performance. Ammonia on the other hand, will do permanent damage, and this damage will steadily get worse and worse.

Ammonia as a hydrogen carrier can easily be compared to methanol. If it were, the following points would be made:

- the production methods and costs are similar;
- the product hydrogen per litre of carrier is slightly better;
- ammonia is far harder to store, handle and transport;
- ammonia is more dangerous and toxic;
- the process of extracting the hydrogen is more complex;

- the reformer operates at very high temperatures, making integration into small fuel cell systems much more difficult than for methanol;
- the product gas is difficult to use with any type of fuel cell other than alkaline.

The conclusion must be that the use of ammonia as a hydrogen carrier is going to be confined to only the most unusual circumstances.

5.4.6 Storage methods compared

We have looked at a range of hydrogen storage methods. In Section 5.3 we looked at fairly simply 'hydrogen in, hydrogen out' systems. In Section 5.4 we looked at some more complex systems involving the use of hydrogen-rich chemicals that can be used as carriers.

None of the methods is without major problems. Table 5.12 compares the systems that are currently feasible in relation to gravimetric and volumetric effectiveness. Together with the summary comments, this should enable the designer to choose the least difficult alternative. It is worth noting that the method with the worst figures (storage in high pressure cylinders) is actually the most widely used. This is because it is so simple and straightforward. The figures also show why methanol is such a promising candidate for the future.

Table 5.12 Data for comparing methods of storing hydrogen fuel. The figures include the associated equipment, e.g. tanks for liquid hydrogen, or reformers for methanol

Method	Gravimetric storage efficiency, % mass H_2	Volumetric mass (in kg) of H_2 per litre	Comments
High pressure in cylinders	0.7–3	0.015	'Cheap and cheerful', widely used
Metal hydride	0.65	0.028	Suitable for small systems
Cryogenic liquid	14.2	0.040	Widely used for bulk storage
Methanol	6.9	0.055	Low cost chemical, potentially useful in a wide range of systems
Sodium hydride pellets	2.2	0.02	Problem of disposing of spent solution
$NaBH_4$ solution in water	3.35	0.036	Very expensive to run

References

Bossel U.G. (1999) Portable fuel cell battery charger with integrated hydrogen generator. Proceedings of the European Fuel Cell Forum Portable Fuel Cells Conference, Lucerne, pp. 79–84.

Brown H.C. (1992) The Discovery of New Continents of Chemistry, lecture given in 1992, available at www.chem.purdue.edu/hcbrown/Lecture.htm

Edwards N., Ellis S.R., Frost J.C., Golunski S.E., van Keulen A.N.J., Lindewald N.G. and Reinkingh J.G. (1998) Onboard hydrogen generation for transport applications: The HotSpot™ methanol processor. *Journal of Power Sources*, Vol. 781, pp. 123–128.

Eliasson B. and Bossel U. (2002) The future of the hydrogen economy, bright or bleak? *The Fuel Cell World*, EFCF Conference Proceedings, pp. 367–382.

Ellis S.R., Golunski S.E. and Petch M.I. (2001) Hotspot processor for reformulated gasoline. ETSU report no. F/02/00143/REP. DTI/Pub URN 01/958.

Faleschini G., Hacker V., Muhr M., Kordesch K. and Aronsson R. (2000) Ammonia for high density hydrogen storage, published at www.electricauto.com/HighDensity_STOR.htm

Hord J. (1978) Is hydrogen a safe fuel? *International Journal of Hydrogen Energy*, Vol. 3, pp. 157–176.

Joensen F. and Rostrup-Nielsen J.R. (2002) Conversion of hydrocarbons and alcohols for fuel cells. *Journal of Power Sources*, 105 (2), pp. 195–201.

Kahrom H. (1999) Clean hydrogen for portable fuel cells. Proceedings of the European Fuel Cell Forum Portable Fuel Cells Conference, Lucerne, pp. 159–170.

Kalhammer F.R., Prokopius P.R., Roan V. and Voecks G.E. (1998) Status and prospects of fuel cells as automobile engines, report prepared for the State of California Air Resources Board.

Kaye I.W., Bloomfield D.P. (1998) Portable ammonia powered fuel cell. *Conference Power Sources*, Cherry Hill, pp. 759–766.

Kordesch K., Hacker V., Gsellmann J., Cifrain M., Faleschini G., Enzinger P., Ortner M., Muhr M. and Aronsson R. (1999) Alkaline Fuel Cell Applications. Proceedings of the 3rd International Fuel Cell Conference, Nagoya, Japan, 1999.

Koschany P. (2001) Hydrogen sources integrated in fuel cells, *The Fuel Cell Home*, EFCF Conference Proceedings, pp. 293–298.

Larminie J. (2002) Sodium Borohydride: is this the answer for fuelling small fuel cells? *The Fuel Cell World*, EFCF Conference Proceedings, pp. 60–67.

Larminie J. and Dicks A. (2003) *Fuel Cell Systems Explained*, 2nd Edn Wiley, Chichester.

Reister D. and Strobl W. (1992) Current development and outlook for the hydrogen fuelled car, in *Hydrogen Energy Progress IX*, pp. 1202–1215.

Teagan W.P., Bentley J. and Barnett B. (1998) Cost implications of fuel cells for transport applications: fuel processing options. *Journal of Power Sources*, Vol. 71, pp. 80–85.

Zieger J. (1994) HYPASSE – Hydrogen powered automobiles using seasonal and weekly surplus of electricity, in *Hydrogen Energy Progress X*, pp. 1427–1437.

Hord (1978), Reister and Strobl (1992) and Zieger (1994) are reprinted in Norbeck J.M. *et al.* (1996) *Hydrogen Fuel for Surface Transportation*, pub. Society of Automotive Engineers.

6

Electric Machines and their Controllers

Clearly, electric motors are a key component of an electric vehicle, and in this chapter we consider the main types of motor used. The more advanced modern electrical machines need fairly complex controllers, so these are also described in this chapter.

We start, in Section 6.1, with an explanation of the simplest types of direct current (DC) motor, that can run with hardly any electronic control. In Section 6.2 we cover the basics of the power electronics that are needed to operate the more advanced motors frequently used now. Then in Section 6.3 we consider the different types of these more sophisticated motors, such as 'switched reluctance', 'brushless', and the tried and tested induction motor. There are issues that apply to all motors when it comes to their selection and use, and these aspects, such as cooling, efficiency, size, and mass are considered in Section 6.4. Finally, in Section 6.5, we consider the special factors that apply to electrical machines in hybrid electric vehicles.

6.1 The 'Brushed' DC Electric Motor

6.1.1 Operation of the basic DC motor

Electric vehicles use what can seem a bewildering range of different types of electric motor. However, the simplest form of electric motor, at least to understand, is the 'brushed' DC motor. This type of motor is very widely used in applications such as portable tools, toys, electrically operated windows in cars, and small domestic appliances such as hair dryers, even if they are AC mains powered.[1] However, they are also still used as traction motors, although the other types of motor considered later in this chapter are becoming more common for this application. The brushed DC motor is a good starting point because, as well as being widely used, most of the important issues in electric motor control can be more easily explained with reference to this type of motor.

---

[1] In this case the appliance will also have a small rectifier.

Electric Vehicle Technology Explained James Larminie and John Lowry
© 2003 John Wiley & Sons, Ltd ISBN: 0-470-85163-5

The classical DC electric motor is shown in Figure 6.1. It is a DC motor, equipped with permanent magnets and brushes. This simplified motor has one coil, and the current passing through the wire near the magnet causes a force to be generated in the coil. The current flows through brush X, commutator half ring A, round the coil, and out through the other commutator half ring B and brush Y (XABY). On one side (as shown in the diagram) the force is upwards, and in the other the force is downwards, because the current is flowing back towards the brushes and commutator. The two forces cause the coil to turn. The coil turns with the commutator, and once the wires are clear of the magnet the momentum carries it on round until the half rings of the commutator connect with the brushes again. When this happens the current is flowing in the same direction relative to the magnets, and hence the forces are in the same direction, continuing to turn the motor as before. However, the current will now be flowing through brush X, half ring B, round the coil to A and out through Y, so the current will be flowing in the opposite direction through the coil (XBAY).

The commutator action ensures that the current in the coil keeps changing direction, so that the force is in the same direction, even through the coil has moved.

Clearly, in a real DC motor there are many refinements over the arrangement of Figure 6.1. The most important of these are as follows.

- The rotating wire coil, often called the armature, is wound round a piece of iron, so that the magnetic field of the magnets does not have to cross a large air gap, which would weaken the magnetic field.
- More than one coil will be used, so that a current-carrying wire is near the magnets for a higher proportion of the time. This means that the commutator does not consist of two half rings (as in Figure 6.1) but several segments, two segments for each coil.

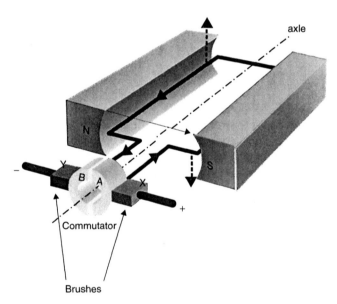

Figure 6.1 Diagram to explain the operation of the simple permanent magnet DC motor

- Each coil will consist of several wires, so that the torque is increased (more wires, more force).
- More than one pair of magnets may be used, to further increase the turning force.

Figure 6.2(a) is the cross-section diagram of a DC motor several steps nearer reality than that of Figure 6.1. Since we are in cross-section, the electric current is flowing in the wires either up out of the page, or down into the page. Figure 6.2(b) shows the convention used when using such diagrams. It can be seen that most of the wires are both carrying a current and in a magnetic field. Furthermore, all the wires are turning the motor in the same direction.

6.1.2 Torque speed characteristics

If a wire in an electric motor has a length l metres, carries a current I amps, and is in a magnetic field of strength B Wb.m^{-2}, then the force on the wire is:

$$F = BIl \tag{6.1}$$

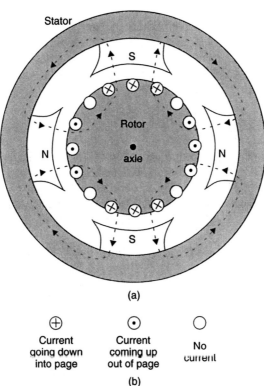

Figure 6.2 (a) Cross-section through a four-pole DC motor. The dotted lines shows the magnetic flux. The motor torque is clockwise. (b) shows the convention used to indicate the direction of current flow in wires drawn in cross-section

If the radius of the coil is r, and the armature consists of n turns, then the motor torque T is given by the equation:

$$T = 2nrBIl \quad (6.2)$$

The term $2Blr = B \times$ area can be replaced by Φ, the total flux passing through the coil. This gives:

$$T = n\Phi I \quad (6.3)$$

However, this is the peak torque, when the coil is fully in the flux, which is perfectly radial. In practice this will not always be so. Also, it does not take into account the fact that there may be more than one pair of magnetic poles, as in Figure 6.2. So we use a constant K_m, known as the motor constant, to connect the average torque with the current and the magnetic flux. The value of K_m clearly depends on the number of turns in each coil, but also on the number of pole pairs, and other aspects of motor design. Thus we have:

$$T = K_m \Phi I \quad (6.4)$$

We thus see that the motor torque is directly proportional to the rotor (also called armature) current I. However, what controls this current? Clearly it depends on the supply voltage E_S to the motor. It will also depend on the electrical resistance of the armature coil R_a. As the motor turns the armature will be moving in a magnetic field. This means it will be working as a generator or dynamo. If we consider the basic machine of Figure 6.1, and consider one side of the coil, the voltage generated is expressed by the basic equation:

$$E_b = Blv \quad (6.5)$$

This equation is the generator form of equation (6.1). The voltage generated is usually called the back EMF, hence the symbol E_b. It depends on the velocity v of the wire moving through the magnetic field. To develop this further, the velocity of the wire moving in the magnetic field depends on ω the angular velocity and r the radius according to the simple equation $v = r\omega$. Also, the armature has two sides, so equation (6.5) becomes:

$$E_b = 2Blr\omega$$

However, as there are many turns, we have:

$$E_b = 2nrBl\omega$$

This equation should be compared with equation (6.2). By similar reasoning we simplify it to an equation like (6.4). Since it is the same motor, the constant K_m can be used again, and it obviously has the same value. The equation gives the voltage or back EMF generated by the dynamo effect of the motor as it turns.

$$E_b = K_m \Phi \omega \quad (6.6)$$

Electric Machines and their Controllers

This voltage opposes the supply voltage E_s and acts to reduce the current in the motor. The net voltage across the armature is the difference between the supply voltage E_s and the back EMF E_b. The armature current is thus:

$$I = \frac{V}{R_a} = \frac{E_s - E_b}{R_a} = \frac{E_s}{R_a} - \frac{K_m \Phi}{R_a} \omega$$

This equation shows that the current falls with increasing angular speed. We can substitute it into equation (6.4) to get the equation connecting the torque and the rotational speed.

$$T = \frac{K_m \Phi E_s}{R_a} - \frac{(K_m \Phi)^2}{R_a} \omega \tag{6.7}$$

This important equation shows that the torque from this type of motor has a maximum value at zero speed, when stalled, and it then falls steadily with increasing speed. In this analysis we have ignored the losses in the form of torque needed to overcome friction in bearings, and at the commutator, and windage losses. This torque is generally assumed to be constant, which means the general form of equation (6.7) still holds true, and gives the characteristic graph of Figure 6.3.

The simple linear relationship between speed and torque, implied by equation (6.7), is replicated in practice for this type of constant magnetic flux DC motor. However, except in the case of very small motors, the low speed torque is reduced, either by the electronic controller, or by the internal resistance of the battery supplying the motor. Otherwise the currents would be extremely high, and would damage the motor. Let us take an example. A popular motor used on small electric vehicles is the 'Lynch' type machine, an example

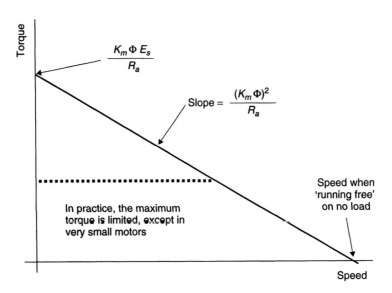

Figure 6.3 Torque/speed graph for a brushed DC motor

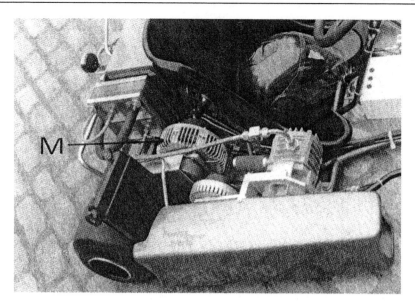

Figure 6.4 Small (10 kW) DC 'Lynch' type DC motor, which is labelled **M**. This go-kart is fuel cell powered. The unit in front of the motor is the air pump, which is driven by its own motor, a smaller version of the traction motor

of which is shown in Figure 6.4. A typical motor of this type[2] might have the following data given in its specification:

- motor speed = 70 rpm/V
- armature resistance = 0.016 Ω

The motor speed information connects with equation (6.7), and refers to the no load speed. Equation (6.7) can be rearranged to:

$$\omega = \frac{E}{K_m \Phi} \text{ rad.s}^{-1} = \frac{60}{2\pi K_m \Phi} E \text{ r.p.m.}$$

So in this case we can say that:

$$\frac{60}{2\pi K_m \Phi} = 70 \Rightarrow K_m \Phi = \frac{60}{2\pi \times 70} = 0.136$$

If this motor were to be run off a fixed 24 V supply equation (6.7) for this motor would be:

$$T = 205 - 1.16\omega \tag{6.8}$$

---

[2] The data given is for a 1998 model of a Lynch disc armature 'type 200' DC motor.

since R_a is given as $0.016\,\Omega$. However, this would mean an initial, zero speed, torque of 205 Nm. This is a huge figure, but may not seem impossibly large until the current is calculated. At zero speed there is no back EMF, and so only this armature resistance R_a opposes the 24 V supply, and so the current would be:

$$I\frac{V}{R} = \frac{E_s}{R_a} = \frac{24}{0.016} = 1500\,\text{A}$$

This is clearly far too large a current. The stated limit on current is 250 A, or 350 A for up to 5 s. We can use this information, and equation (6.4) to establish the maximum torque as:

$$T = K_m \Phi I = 0.136 \times 250 = 34\,\text{Nm} \qquad (6.9)$$

Equations (6.8), modified by equation (6.9) to give a maximum torque, is typical of the characteristic equations of this type of motor. The maximum power is about 5 kW.

6.1.3 Controlling the brushed DC motor

Figure 6.3 and equation (6.7) show us that the brushed DC motor can be very easily controlled. If the supply voltage E_S is reduced, then the maximum torque falls in proportion, and the slope of the torque/speed graph is unchanged. In other words any torque and speed can be achieved below the maximum values. We will see in Section 6.2 that the supply voltage can be controlled simply and efficiently, so this is a good way of controlling this type of motor.

However, reducing the supply voltage is not the only way of controlling this type of motor. In some cases we can also achieve control by changing the magnetic flux Φ. This is possible if coils rather than permanent magnets provide the magnetic field. If the magnetic flux is reduced then the maximum torque falls, but the slope of the torque/speed graph becomes flatter. Figure 6.5 illustrates this. Thus the motor can be made to work at a wide range of torque and speed. This method is sometimes better than simply using voltage control, especially at high speed/low torque operation, which is quite common in electric vehicles cruising near their maximum speed. The reason for this is that the iron losses to be discussed in Section 6.1.5 below, and which are associated with high speeds and strong magnetic fields, can be substantially reduced.

So the brushed DC motor is very flexible as to control method, especially if the magnetic flux Φ can be varied. This leads us to the next section, where the provision of the magnetic flux is described.

6.1.4 Providing the magnetic field for DC motors

In Figures 6.1 and 6.2 the magnetic field needed to make the motor turn is provided by permanent magnets. However, this is not the only way this can be done. It is possible to use coils, through which a current is passed, to produce the magnetic field. These *field windings* are placed in the stator of the electric motor.

An advantage of using electro-magnets to provide the magnetic field is that the magnetic field strength Φ can be changed, by changing the current. A further advantage is that it

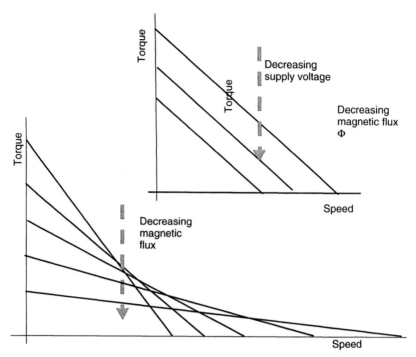

Figure 6.5 How changing the supply voltage and the magnetic field strength affects the torque speed characteristic of the DC motor

is a cheaper way of producing a strong magnetic field, though this is becoming less and less of a factor as the production of permanent magnets improves. The main disadvantage is that the field windings consume electric current, and generate heat; thus it seems that the motor is almost bound to be less efficient. In practice the extra control of magnetic field can often result in *more* efficient operation of the motor, as the iron losses to be discussed in the next section can be reduced. The result is that brushed DC motors with field windings are still often used in electric vehicles.

There are three classical types of brushed DC motor with field windings, as shown in Figure 6.6. However, only one need concern us here. The behaviour of the 'series' and 'shunt' motors is considered in books on basic electrical engineering. However, they do not give the control of speed and torque that is required in an electric vehicle, and the only serious contender is the 'separately excited' motor, as in Figure 6.6(c).

The shunt (or parallel) wound motor of Figure 6.6(a) is particularly difficult to control, as reducing the supply voltage also results in a weakened magnetic field, thus reducing the back EMF, and tending to increase the speed. A reduction in supply voltage can in some circumstances have very little effect on the speed. The particular advantage of the series motor of Figure 6.6(b) is that the torque is very high at low speeds, and falls off rapidly as the speed rises. This is useful in certain applications, for example the starter motor of internal combustion engines, but it is not what is usually required in traction applications.

Electric Machines and their Controllers

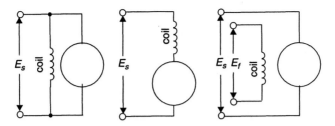

Figure 6.6 Three standard methods of supplying current to a coil providing the magnetic field for brushed DC motors

The separately excited motor of Figure 6.6(c) allows us to have independent control of both the magnetic flux Φ (by controlling the voltage on the field winding E_f) and also the supply voltage E_S. This allows the required torque at any required angular speed to be set with great flexibility. It allows both the control methods of Figure 6.5 to be used, reducing armature supply voltage E_S or reducing the magnetic flux Φ.

For these reasons the separately excited brushed DC motor is quite widely used as the traction motor in electric vehicles. In the case of the many smaller motors that are found on any vehicle, the magnetic field is nearly always provided by permanent magnets. This makes for a motor that is simpler and cheaper to manufacture. Such permanent magnet motors are also sometimes used as traction motors.

6.1.5 DC motor efficiency

The major sources of loss in the brushed DC electric motor are the same as for all types of electric motor, and can be divided into four main types, as follows.

Firstly there are the **copper losses**. These are caused by the electrical resistance of the wires (and brushes) of the motor. This causes heating, and some of the electrical energy supplied is turned into heat energy rather than electrical work. The heating effect of an electrical current is proportional to the square of the current:

$$P = I^2 R$$

However, we know from equations (6.3) and (6.4) that the current is proportional to the torque T provided by the motor, so we can say that:

$$\text{Copper losses} = k_c T^2 \tag{6.10}$$

where k_c is a constant depending on the resistance of the brushes and the coil, and also the magnetic flux Φ. These copper losses are probably the most straightforward to understand and, especially in smaller motors, they are the largest cause of inefficiency.

The **second** major source of losses is called **iron losses**, because they are caused by magnetic effects in the iron of the motor, particularly in the rotor. There are two main causes of these iron losses, but to understand both it must be understood that the magnetic field in the rotor is continually changing. Imagine a small ant clinging onto the edge of the rotor of Figure 6.2. If the rotor turns round one turn then this ant will pass a north pole, then a south pole, and then a north pole, and so on. As the rotor rotates the magnetic field supplied by the magnets may be unchanged, but that seen by the turning rotor (or the ant clinging to it) is always changing. Any one piece of iron on the rotor is thus effectively in an ever-changing magnetic field. This causes two types of loss. The first is called 'hysteresis' loss, and is the energy required to continually magnetise and demagnetise the iron, aligning and re-aligning the magnetic dipoles of the iron. In a good magnetically soft iron this should be very small, but will not be zero. The second iron loss results from the fact that the changing magnetic field will generate a current in the iron, by the normal methods of electromagnetic induction. This current will result in heating of the iron. Because these currents just flow around and within the iron rotor they are called 'eddy currents'. These eddy currents are minimised by making the iron rotor, not out of one piece, but using thin sheets all bolted or glued together. Each sheet is separated from its neighbour by a layer of paint. This greatly reduces the eddy currents by effectively increasing the electrical resistance of the iron.

It should be clear that these iron losses are proportional to the *frequency* with which that magnetic field changes; a higher frequency results in more magnetising and demagnetising, and hence more hysteresis losses. Higher frequency also results in a greater rate of change of flux, and hence a greater induced eddy currents. However, the rate of change of magnetic flux is directly proportional to the speed of the rotor; to how quickly it is turning. We can thus say that:

$$\text{Iron losses} = k_i \omega \tag{6.11}$$

where k_i is a constant. In fact, it will not really be constant, as its value will be affected by the magnetic field strength, among other non-constant factors. However, a single value can usually be found which gives a good indication of iron losses. The degree to which we can say k_i is constant depends on the way the magnetic field is provided; it is more constant in the case of the permanent magnet motor than the separately excited.

The **third** category of loss is that due to **friction and windage**. There will of course be a friction torque in the bearings and brushes of the motor. The rotor will also have a wind resistance, which might be quite large if a fan is fitted to the rotor for cooling. The friction force will normally be more or less constant. However, the wind resistance force will increase with the square of the speed. To get at the power associated with these forces, we must multiply by the speed, as:

$$\text{power} = \text{torque} \times \text{angular speed}$$

the power involved in these forces will then be:

$$\text{friction power} = T_f \omega \quad \text{and} \quad \text{windage power} = k_w \omega^3 \tag{6.12}$$

where T_f is the friction torque, and k_w is a constant depending mainly on the size and shape of the rotor, and whether or not a cooling fan is fitted.

Finally, we address those **constant losses** that occur even if the motor is totally stationary, and vary neither with speed or torque. In the case of the separately excited motor

Electric Machines and their Controllers

these are definitely not negligible, as current (and hence power) must be supplied to the coil providing the magnetic field. In the other types of motor to be described in the sections that follow, power is needed for the electronic control circuits that operate at all times. The only type of motor for which this type of loss could be zero is the permanent magnet motor with brushes. The letter C is used to designate these losses.

It is useful to bring together all these different losses into a single equation that allows us to model and predict the losses in a motor. When we do this it helps to combine the terms for the iron losses and the friction losses, as both are proportional to motor speed. Although we have done this for the brushed DC motor, it is important to note that this equation is true, to a good approximation, for all types of motor, including the more sophisticated types to be described in later section.

If we combine equations (6.10), (6.11) and (6.12), we have:

$$\text{total losses} = k_c T^2 + k_i \omega + k_w \omega^3 + C \tag{6.13}$$

However, it is usually the motor efficiency η_m that we want. This is found as follows:

$$\eta_m = \frac{\text{output power}}{\text{input power}}$$

$$\eta_m = \frac{\text{output power}}{\text{output power} + \text{losses}} = \frac{T\omega}{T\omega + k_c T^2 + k_i \omega + k_w \omega^3 + C} \tag{6.14}$$

This equation will be very useful when we come to model the performance of electrical vehicles in Chapter 7. Suitable values for the constants in this equation can usually be found by experimentation, or by regression using measured values of efficiency. For example, typical values for a permanent magnet motor of the 'Lynch' type that we were considering in Section 6.1.2, that might be fitted to an electric scooter are as follows:

$$k_c = 0.8 \quad k_i = 0.1 \quad k_w = 10^{-5} \quad C = 20$$

It is useful to plot the values of efficiency on a torque speed graph, giving what is sometimes known as an efficiency map for the motor, which gives an idea of the efficiency at any possible operating condition. Such a chart is shown in Figure 6.7. MATLAB is an excellent program for producing plots of this type, and in Appendix 1 we have included the script file used to produce this graph.

6.1.6 Motor losses and motor size

While it is obvious that the losses in a motor affect its efficiency, it is not so obvious that the losses also have a crucial impact on the maximum power that can be obtained from a motor of any given size.

Consider a brushed motor of the type we have described in this section. The power produced could be increased by increasing the supply voltage, and thus the torque, as per Figure 6.6. Clearly, there must be a limit to this, the power cannot be increased to infinity. One might suppose the limiting factor is the voltage at which the insulation around the

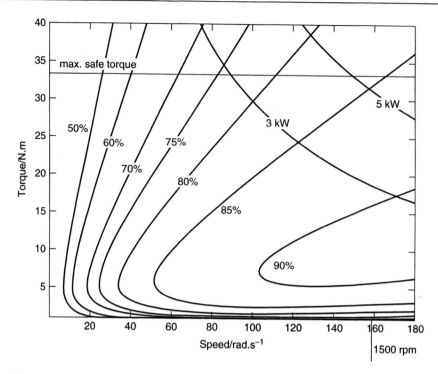

Figure 6.7 Efficiency map for a typical permanent magnet DC motor, with brushes

copper wire breaks down, or some such point. However, that is not the case. The limit is in fact temperature-related. Above a certain power the heat generated as a result of the losses, as given by equation (6.13), become too large to be conducted, convected and radiated away, and the motor overheats.

An important result of this is that the key electric motor parameters of *power density* and *specific power*, being the power per unit volume and the power per kilogram mass, are not controlled by electrical factors so much as how effectively the waste heat can be removed from the motor.[3]

This leads to a very important disadvantage of the classical brushed DC motor. In this type of motor virtually all the losses occur in the rotor at the centre of the motor. This means that the heat generated is much more difficult to remove. In the motors to be considered in later sections the great majority of the losses occur on the stator, the stationary outer part of the motor. Here they can much more easily be removed. Even if we stick with air-cooling it can be done more effectively, but in larger motors liquid cooling can by used to achieve even higher power density.

This issue of motor power being limited by the problem of heat removal also explains another important feature of electric motors. This is that they can safely be driven well in excess of their rated power for short periods. For example, if we take a motor that has a

---

[3] Though obviously the losses are affected by electrical factors, such as coil resistance.

rated power of 5 kW, this means that if it is run at this power for about 30 minutes, it will settle down to a temperature of about 80°C, which is safe and will do it no harm. However, being fairly large and heavy, a motor will take some time to heat up. If it is at, say, 50°C, we can run it in excess of 5 kW, and its temperature will begin to increase quite rapidly. However, if we do not do this for more than about 1 minute, then the temperature will not have time to rise to a dangerous value. Clearly this must not be overdone, otherwise local heating could cause damage, nor can it be done for too long, as a dangerous temperature will be reached. Nevertheless, in electric vehicles this is particularly useful, as the higher powers are often only required for short time intervals, such as when accelerating.

6.1.7 Electric motors as brakes

The fact that an electric motor can be used to convert kinetic energy back into electrical energy is an important feature of electric vehicles. How this works is easiest to understand in the case of the classical DC motor with brushes, but the broad principles apply to all motor types.

Consider Figure 6.8. A DC motor is connected to a battery of negligible internal resistance, and voltage E_S. It reaches a steady state, providing a torque T at a speed ω. These variables will be connected by equation (6.7). Suppose the switch S is now moved over to the right. The motor will continue to move at the same angular speed. This will cause a voltage to be generated, as given by equation (6.6). This voltage will be applied to the resistor R_L, as in Figure 6.8, with the current further limited by the resistance of the rotor coil (armature). The result is that the current will be given by the formula:

$$I = \frac{K_m \Phi \omega}{R_a + R_L} \quad (6.15)$$

This current will be flowing out of the motor, and will result in a negative torque. The value of this torque will still be given by the torque equation produced earlier as equation (6.4). So, the negative torque, which will slow the motor down, will be given by the equation:

$$T = -\frac{(K_m \Phi)^2 \omega}{R_a + R_L} \quad (6.16)$$

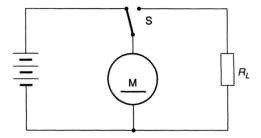

Figure 6.8 Motor circuit with resistor to be used for dynamic braking

We thus have a negative torque, whose value can be controlled by changing the resistance R_L. The value of this torque declines as the speed ω decreases. So, if R_L is constant we might expect the speed to decline in an exponential way to zero.

This way of slowing down an electric motor, using a resistor, is known as dynamic braking. Note that all the kinetic energy of the motor (and the vehicle connected to it) is ultimately converted into heat, just like normal friction brakes. However, we do have control of where the heat is produced, which can be useful. We also have the potential of an elegant method of controlling the braking torque. Nevertheless, the advance over normal friction brakes is not very great, and it would be much better if the electrical energy produced by the motor could be stored in a battery or capacitor.

If the resistor of Figure 6.8 were replaced by a battery, then we would have a system known as regenerative braking. However, the simple connection of a battery to the motor is not practical. Suppose the voltage of the battery is V_b, and the motor is turning at speed ω, then the current that will flow out of the motor will be given by the equation:

$$I = \frac{V}{R} = \frac{K_m \Phi \omega - V_b}{R_a} \tag{6.17}$$

The slowing down torque will be proportional to this current. Once the value of ω reaches the value where the voltage generated by the motor ($= K_m \Phi \omega$) reaches the battery voltage, then there will be no more braking effect. Unless the battery voltage is very low, then this will happen quite soon. If the battery voltage is low, then it will be difficult to use the energy stored in it, and the braking effect might well be far too strong at high speeds, with the current given by equation (6.17) being impracticably large.

The solution lies in a voltage converter circuit as in Figure 6.9. The converter unit, know as a DC/DC converter, draws a current from the motor I_m, which will occur at a voltage V_m. This voltage V_m will change with motor (and hence vehicle) speed. The current I_m will change with the desired braking torque. The DC/DC converter will take this electrical power ($= V_m \times I_m$) and put it out at an increased voltage (and reduced current) so that it matches the rechargeable battery or capacitor that is storing the energy. The battery might well be the same battery that provided the electricity to make the motor go in the first place. The key point is that the motor voltage might be considerably lower than the battery voltage, but it can still be providing charge to the battery.

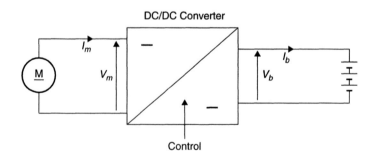

Figure 6.9 Regenerative braking of a DC motor

Electric Machines and their Controllers 155

Such a converter circuit sounds as if it is a little 'too good to be true', and the possibility of it remote. It seems to be like getting water to flow uphill. However, such circuits are by quite possible with modern power electronics. We are not producing power, we are exchanging a low voltage and high current for a higher voltage and lower current. It is like a transformer in AC circuits, with the added facility of being able to continuously vary the ratio of the input and output voltages.

Although voltage converter circuits doing what is described above can be made, they are not 100% efficient. Some of the electrical power from the braking motor will be lost. We can say that:

$$V_b \times I_b = \eta_c \times V_m \times I_m \qquad (6.18)$$

where η_c is the efficiency of the converter circuit.

We have thus seen that a motor can be used to provide a controllable torque over a range of speeds. The motor can also be used as a brake, with the energy stored in a battery or capacitor. To have this range of control we need power electronics circuits that can control the voltages produced. The operation of these circuits is considered in the section that follows.

6.2 DC Regulation and Voltage Conversion

6.2.1 Switching devices

The voltage from all sources of electrical power varies with time, temperature, and many other factors, especially current. Fuel cells, for example, are particularly badly regulated, and it will always be necessary to control the output voltage so that its only varies between set boundaries. Battery voltage is actually quite well regulated, but frequently we will want to change the voltage to a lower or higher value, usually to control the speed of a motor. We saw in the last section that if an electric motor is to be used in regenerative braking we need to be able to boost the voltage (and reduce the current) in a continuously variable way.

A good example to illustrate the variable voltage from a fuel cell system is given in Figure 6.10. It summarises some data from a real 250 kW fuel cell used to drive a bus (Spiegel *et al.* 1999). The voltage varies from about 400 to over 750 V, and we also see that the voltage can have different values at the same current. This is because, as well as current, the voltage also depends on temperature, air pressure, and on whether or not the compressor has got up to speed, among other factors.

Most electronic and electrical equipment requires a fairly constant voltage. This can be achieved by dropping the voltage down to a fixed value below the operating range of the fuel cell or battery, or boosting it up to a fixed value. In other cases we want to produce a variable voltage (e.g. for a motor) from the more-or-less fixed voltage of a battery. Whatever change is required, it is done using 'switching' or 'chopping' circuits, which are described below. These circuits, as well as the inverters and motor controllers to be described in later sections, use *electronic switches*.

As far as the user is concerned, the particular type of electronic switch used does not matter greatly, but we should briefly describe the main types used, so that the reader

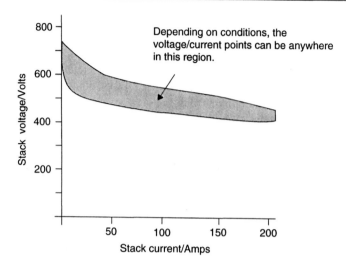

Figure 6.10 Graph summarising some data from a real 250 kW fuel cell used to power a bus (Derived from data in Spiegel *et al.* (1999).)

Table 6.1 Key data for the main types of electronic switch used in modern power electronic equipment

Type	Thyristor	MOSFET	IGBT
symbol			
Max. voltage (V)	4500	1000	1700
Max. current (A)	4000	50	600
Switching time (μs)	10–25	0.3–0.5	1–4

has some understanding of their advantages and disadvantages. Table 6.1 shows the main characteristics of the most commonly used types.

The metal oxide semiconductor field effect transistor (MOSFET) is turned on by applying a voltage, usually between 5 and 10 V, to the gate. When 'on', the resistance between the drain (d) and source (s) is very low. The power required to ensure a very low resistance is small, as the current into the gate is low. However, the gate does have a considerable capacitance, so special drive circuits are usually used. The current path behaves like a resistor, whose 'on' value is $R_{DS_{ON}}$. The value of $R_{DS_{ON}}$ for a MOSFET used in voltage regulation circuits can be as low as about 0.01 Ω. However, such low values are only possible with devices that can switch low voltages, in the region of 50 V. Devices which can switch higher voltages have values of $R_{DS_{ON}}$ of about 0.1 Ω, which causes higher losses. MOSFETs are widely used in low voltage systems of power less than about 1 kW.

The insulated gate bipolar transistor (IGBT) is essentially an integrated circuit combining a conventional bipolar transistor and a MOSFET, and it has the advantages of both. They require a fairly low voltage, with negligible current at the gate to turn on. The main current flow is from the collector to the emitter, and this path has the characteristics of a p-n junction. This means that the voltage does not rise much above 0.6 V at all current within the rating of the device. This makes it the preferred choice for systems where the current is greater than about 50 A. They can also be made to withstand higher voltages. The longer switching times compared to the MOSFET, as given in Table 6.1, are a disadvantage in lower power systems. However, the IGBT is now almost universally the electronic switch of choice in systems from 1 kW up to several hundred kW, with the upper limit rising each year.

The thyristor has been the electronic switch most commonly used in power electronics. Unlike the MOSFET and IGBT the thyristor can only be used as an electronic switch, it has no other applications. The transition from the blocking to the conducting state is triggered by a pulse of current into the gate. The device then remains in the conducting state until the current flowing through it falls to zero. This feature makes them particularly useful in circuits for rectifying AC, where they are still widely used. However, various variants of the thyristor, particularly the gate-turn-off, or GTO thyristor, can be switched off, even while a current is flowing, by the application of a negative current pulse to the gate.

Despite the fact that the switching is achieved by just a pulse of current, the energy needed to effect the switching is much greater than for the MOSFET or the IGBT. Furthermore, the switching times are markedly longer. The only advantage of the thyristor (in its various forms) for DC switching is that higher currents and voltages can be switched. However, the maximum power of IGBTs is now so high that this is very unlikely to be an issue in electric vehicle systems, which are usually below 1MW in power.[4]

Ultimately the component used for the electronic switch is not of great importance. As a result the circuit symbol used is often the 'device-independent' symbol shown in Figure 6.11. In use, it is essential that the switch moves as quickly as possible from the conducting to the blocking state, or vice versa. No energy is dissipated in the switch while it is open circuit, and only very little when it is fully on; it is while the transition takes place that the product of voltage and current is non-zero, and that power is lost.

6.2.2 Step-down or 'buck' regulators

The 'step-down' or 'buck' switching regulator (or chopper) is shown in Figure 6.12. The essential components are an electronic switch with an associated drive circuit, a diode and

Figure 6.11 Circuit symbol for a voltage operated electronic switch of any type

---

[4] Electric railways locomotives would be an exception to this, but they are outside the scope of this book.

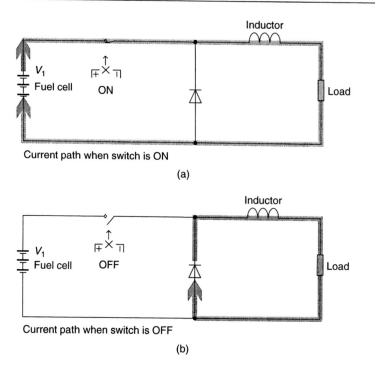

Figure 6.12 Circuit diagram showing the operation of a switch mode step down regulator

an inductor. In Figure 6.12(a) the switch is on, and the current flows through the inductor and the load. The inductor produces a back EMF, making the current gradually rise. The switch is then turned off. The stored energy in the inductor keeps the current flowing through the load, using the diode, as in Figure 6.12(b). The different currents flowing during each part of this on-off cycle are shown in Figure 6.13. The voltage across the load can be further smoothed using capacitors if needed.

If V_1 is the supply voltage, and the 'on' and 'off' times for the electronic switch are t_{ON} and t_{OFF}, then it can be shown that the output voltage V_2 is given by:

$$V_2 = \frac{t_{ON}}{t_{ON} + t_{OFF}} V_1 \qquad (6.19)$$

It is also clear that the ripple depends on the frequency: higher frequency, less ripple. However, each turn-on and turn-off involves the loss of some energy, so the frequency should not be too high. A control circuit is needed to adjust t_{ON} to achieve the desired output voltage; such circuits are readily available from many manufacturers.

The main energy losses in the step-down chopper circuit are:

- switching losses in the electronic switch;
- power lost in the switch while on ($0.6 \times I$ for an IGBT, or $R_{DS_{ON}} \times I^2$ for a MOSFET);

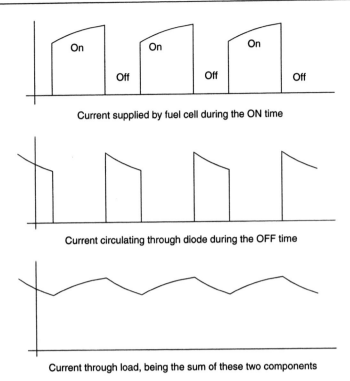

Figure 6.13 Currents in the step down switch mode regulator circuit

- power lost due to the resistance of the inductor;
- losses in the diode, $0.6 \times I$.

In practice all these can be made very low. The efficiency of such a step-down chopper circuit should be over 90%. In higher voltage systems, about 100 V or more, efficiencies as high as 98% are possible.

We should at this point briefly mention the 'linear' regulator circuit. The principle is shown in Figure 6.14. A transistor is used again, but this time it is not switched fully on or fully off. Rather, the gate voltage is adjusted so that its resistance is at the correct value to drop the voltage to the desired value. This resistance will vary continuously, depending on the load current and the supply voltage. This type of circuit is widely used in small electronic systems, but should *never* be used with traction motors. The voltage is dropped by simply converting the surplus voltage into heat. Linear regulators have no place in systems where efficiency is paramount, such as an electric vehicle.

6.2.3 Step-up or 'boost' switching regulator

It is often desirable to step-up or boost a DC voltage, regenerative braking being just one example. This can also be done quite simply and efficiently using switching circuits.

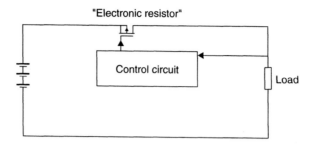

Figure 6.14 Linear regulator circuit

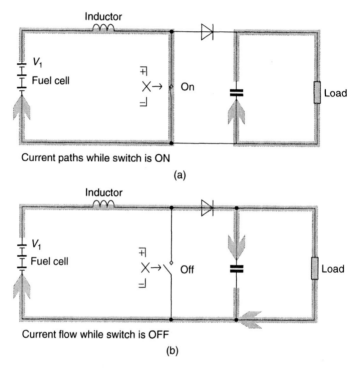

Figure 6.15 Circuit diagram to show the operation of a switch mode boost regulator

The circuit of Figure 6.15 is the basis usually used. We start our explanation by assuming some charge is in the capacitor. In Figure 6.15(a) the switch is on, and an electric current is building up in the inductor. The load is supplied by the capacitor discharging. The diode prevents the charge from the capacitor flowing back through the switch. In Figure 6.15(b) the switch is off. The inductor voltage rises sharply, because the current is falling. As soon as the voltage rises above that of the capacitor (plus about 0.6 V for the diode) the current will flow through the diode, and charge up the capacitor and flow through the load. This will continue so long as there is still energy in the inductor. The switch is then closed again, as in Figure 6.15(a), and the inductor re-energised while the capacitor supplies the load.

Higher voltages are achieved by having the switch off for a short time. It can be shown that for an ideal convertor with no losses:

$$V_2 = \frac{t_{ON} + t_{OFF}}{t_{OFF}} V_1 \qquad (6.20)$$

In practice the output voltage is somewhat less than this. As with the step-down (buck) switcher, control circuits for such boost or step-up switching regulators are readily available from many manufacturers.

The losses in this circuit come from the same sources as for the step-down regulator. However, because the currents through the inductor and switch are higher than the output current, the losses are higher. Also, all the charge passes through the diode this time, and so is subject to the 0.6 V drop and hence energy loss. The result is that the efficiency of these boost regulators is somewhat less than for the buck. Nevertheless, over 80% should normally be obtained, and in systems where the initial voltage is higher (over 100 V), efficiencies of 95% or more are possible.

For the regulation of fuel cell voltages, in cases where a small variation in output voltage can be tolerated, an up-chopper circuit is used at *higher currents only*. This is illustrated in Figure 6.16. At lower currents the voltage is not regulated. The circuit of Figure 6.15 is used, with the switch permanently off. However, the converter starts operating when the fuel cell voltage falls below a set value. Since the voltage shift is quite small, the efficiency would be higher.

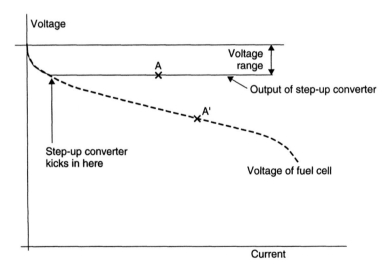

Figure 6.16 Graph of voltage against current for a fuel cell with a step-up chopper circuit that regulates to a voltage a little less than the maximum stack voltage

It should be pointed out that, of course, the current out from a step-up converter is less than the current in. In Figure 6.16, if the fuel cell is operating at point A, the output will be at point A', a higher voltage but a lower current. Also, the system is not entirely loss-free while the converter is not working. The current would all flow through the inductor and the diode, resulting in some loss of energy.

These step-up and step-down switcher or chopper circuits are called DC/DC converters. Complete units, ready made and ruggedly packaged are available as off-the-shelf units in a wide range of powers and input and output voltages. However, when they are used as motor controller circuits, as in the case of electric vehicles, the requirements of having to produce a variable voltage, or a fixed output voltage for a variable input voltage (as in the case when braking a motor using regenerative braking), then such off-the-shelf units will not often be suitable. In such cases special circuits must be designed, and most motors can be supplied with suitable controllers. As we have seen, the circuits required are, in principle, quite simple. The key is to properly control the switching of an electronic switch. This control is usually provided by a microprocessor.

6.2.4 Single-phase inverters

The circuits of the previous two sections are the basis of controlling the classical DC motor. However, the motors to be considered in the next section require alternating current (AC). The circuit that produces AC from DC sources such as batteries and fuel cells is known as an inverter. We will begin with the single phase inverter.

The arrangement of the key components of single phase inverter is shown in Figure 6.17. There are four electronic switches, labelled A, B, C and D, connected in what is called an H-bridge. Across each switch is a diode, whose purpose will become clear later. The load through which the AC is to be driven is represented by a resistor and an inductor.

The basic operation of the inverter is quite simple. First switches A and D are turned on, and a current flows to the right through the load. These two switches are then turned off; at this point we see the need for the diodes. The load will probably have some

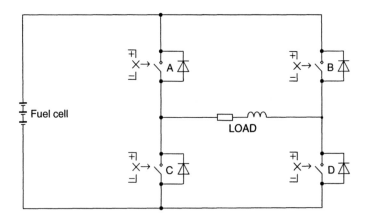

Figure 6.17 H-bridge inverter circuit for producing single phase alternating current

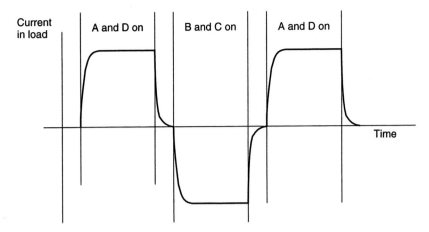

Figure 6.18 Current/time graph for a square wave switched single-phase inverter

inductance, and so the current will not be able to stop immediately, but will continue to flow in the same direction, through the diodes across switches B and C, back into the supply. The switches B and C are then turned on, and a current flows in the opposite direction, to the left. When these switches turn off, the current 'free-wheels' on through the diodes in parallel with switches A and D.

The resulting current waveform is shown in Figure 6.18. The fact that it is very far from a sine wave may be a problem in some cases, which we will consider here.

The difference between a pure sine wave and any other waveform is expressed using the idea of harmonics. These are sinusoidal oscillations of voltage or current whose frequency f_v is a whole number multiple of the fundamental oscillation frequency. It can be shown that *any* periodic waveform of *any* shape can be represented by the addition of harmonics to a fundamental sine wave. The process of finding these harmonics is known as Fourier analysis. For example, it can be shown that a square wave of frequency f can be expressed by the equation:

$$v = \sin(\omega t) - \frac{1}{3}\sin(3\omega t) + \frac{1}{5}\sin(5\omega t) - \frac{1}{7}\sin(7\omega t) + \frac{1}{9}\sin(9\omega t)\ldots$$
$$\text{where} \quad \omega = 2\pi ft$$

So the difference between a voltage or current waveform and a pure sine wave may be expressed in terms of higher frequency harmonics imposed on the fundamental frequency.

In mains-connected equipment these harmonics can cause a wide range of problems, but that is not our concern here. With motors the main problem is that the harmonics can increase the iron losses mentioned back in Section 6.1.5. We saw that these iron losses are proportional to the frequency of the change of the magnetic field. If our AC is being used to produce a changing magnetic field (which it nearly always will be) then the real rate of change, and hence the losses, will be noticeably increased by these higher harmonic frequencies. For this reason the simple switching pattern just described is often not used,

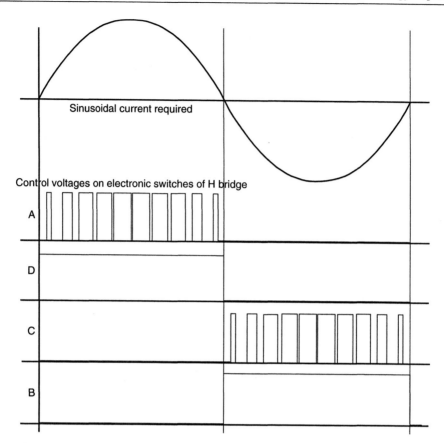

Figure 6.19 Pulse width modulation switching sequence for producing an approximately sinusoidal alternating current from the circuit of Figure 6.17

in favour of a more complex system that produces a more smoothly changing current pattern. This method is known as pulse width modulation.

The principle of pulse width modulation is shown in Figure 6.19. The same circuit as shown in Figure 6.17 is used. In the positive cycle only switch D is on all the time, and switch A is on intermittently. When A is on, current builds up in the load. When A is off, the current continues to flow, because of the load inductance, through switch D and the free-wheeling diode in parallel with switch C, around the bottom right loop of the circuit.

In the negative cycle a similar process occurs, except that switch B is on all the time, and switch C is 'pulsed'. When C is on, current builds in the load; when off, it continues to flow (though declining) through the upper loop in the circuit, and through the diode in parallel with switch A.

The precise shape of the waveform will depend on the nature (resistance, inductance, capacitance) of the load, but a typical half-cycle is shown in Figure 6.20. The waveform is still not a sine wave, but is a lot closer than that of Figure 6.18. Clearly, the more pulses there are in each cycle, the closer will be the wave to a pure sine wave, and the

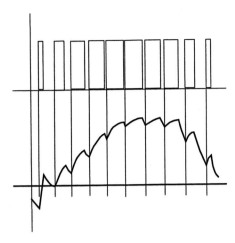

Figure 6.20 Typical voltage/time graph for a pulse modulated inverter

weaker will be the harmonics. Twelve pulses per cycle is a commonly used standard, and generally this gives satisfactory results. In modern circuits the switching pulses are generated by microprocessor circuits.

6.2.5 Three-phase

Most large motors, of the type used in electric vehicles, have three sets of coils rather than just one. For these systems, as well as for regular mains systems, a three-phase AC supply is needed.

This is only a little more complicated than single-phase. The basic circuit is shown in Figure 6.21. Six switches, with free-wheeling diodes, are connected to the three-phase transformer on the right. The way in which these switches are used to generate three similar but out of phase voltages is shown in Figure 6.22. Each cycle can be divided into six steps. The graphs of Figure 6.23 show how the current in each of the three phases changes with time using this simple arrangement. These curves are obviously

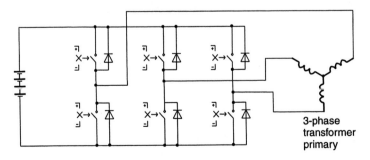

Figure 6.21 Three-phase inverter circuit

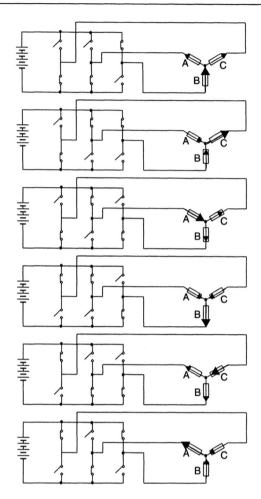

Figure 6.22 Switching pattern to generate three-phase alternating current

far from sine waves. In practice the very simple switching sequence of Figure 6.22 is modified using pulse width modulation, in the same way as for the single-phase inverters described above.

6.3 Brushless Electric Motors

6.3.1 Introduction

In Section 6.1 we described the classical DC electric motor. The brushes of this motor are an obvious problem; there will be friction between the brushes and the commutator, and both will gradually wear away. However, a more serious problem with this type of motor was raised in Section 6.1.5. This is that the heat associated with the losses is generated in

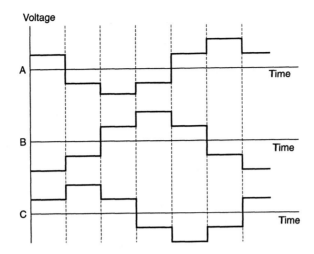

Figure 6.23 Current/time graphs for the simple three-phase AC generation system shown in Figure 6.22, assuming a resistive load. One complete cycle for each phase is shown. Current flowing *out* from the common point is taken as *positive*

the middle of the motor, in the rotor. If the motor could be so arranged that the heat was generated in the outer stator, that would allow the heat to be removed much more easily, and allow smaller motors. If the brushes could be disposed of as well that would be a bonus. In this section we describe three types of motor that are used as traction motors in vehicles that fulfil these requirements.

One of the interesting features of electric motor technology is that there is no clear winner. All three types of motor described here, as well as the brushed DC motor of Section 6.1, are used in current vehicle designs.

6.3.2 The brushless DC motor

The brushless DC motor (BLDC motor) is really an AC motor! The current through it alternates, as we shall see. It is called a brushless DC motor because the alternating current *must* be variable frequency and so derived from a DC supply, and because its speed/torque characteristics are very similar to the ordinary 'with brushes' DC motor. As a result of brushless DC being not an entirely satisfactory name, it is also, very confusingly, given different names by different manufacturers and users. The most common of these is self-synchronous AC motor, but others include variable frequency synchronous motor, permanent magnet synchronous motor, and electronically commutated motor (ECM).

The basis of operation of the BLDC motor is shown in Figure 6.24. Switches direct the direct from a DC source through a coil on the stator. The rotor consists of a permanent magnet. In Figure 6.24(a) the current flows in the direction that magnetises the stator so that the rotor is turned clockwise, as shown. In 6.24(b) the rotor passes between the poles of the stator, and the stator current is switched off. Momentum carries the rotor on, and in 6.24(c) the stator coil is re-energised, but the current and hence the magnetic field, are

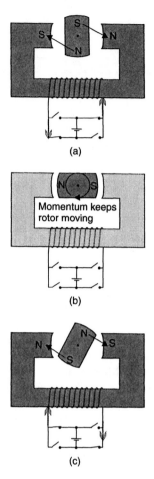

Figure 6.24 Diagram showing the basis of operation of the brushless DC motor

reversed. So the rotor is pulled on round in a clockwise direction. The process continues, with the current in the stator coil alternating.

Obviously, the switching of the current must be synchronised with the position of the rotor. This is done using sensors. These are often Hall effect sensors that use the magnetism of the rotor to sense its position, but optical sensors are also used.

A problem with the simple single coil system of Figure 6.24 is that the torque is very unsteady. This is improved by having three (or more) coils, as in Figure 6.25. In this diagram coil B is energised to turn the motor clockwise. Once the rotor is between the poles of coil B, coil C will be energised, and so on.

The electronic circuit used to drive and control the coil currents is usually called an inverter, and it will be the same as, or very similar to, our universal inverter circuit of Figure 6.21. The main control inputs to the microprocessor will be the position sense signals.

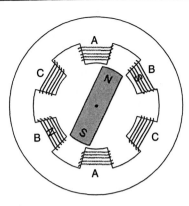

Figure 6.25 Diagram showing an arrangement of three coils on the stator of a BLDC motor

A feature of these BLDC motors is that the torque will reduce as the speed increases. The rotating magnet will generate a back EMF in the coil which it is approaching. This back EMF will be proportional to the speed of rotation, and will reduce the current flowing in the coil. The reduced current will reduce the magnetic field strength, and hence the torque. Eventually the size of the induced back EMF will equal the supply voltage, and at this point the maximum speed has been reached. This behaviour is exactly the same as with the brushed DC motor of Section 6.1.

We should also notice that this type of motor can very simply be used as a generator of electricity, and for regenerative or dynamic braking.

Although the current through the motor coils alternates, there must be a DC supply, which is why these motors are generally classified as DC. They are very widely used in computer equipment to drive the moving parts of disc storage systems and fans. In these small motors the switching circuit is incorporated into the motor with the sensor switches. However, they are also used in higher power applications, with more sophisticated controllers (as of Figure 6.21), which can vary the coil current (and hence torque) and thus produce a very flexible drive system. Some of the most sophisticated electric vehicle drive motors are of this type, and one is shown in Figure 6.26. This is a 100 kW oil-cooled motor, weighing just 21 kg.

These BLDC motors need a strong permanent magnet for the rotor. The advantage of this is that currents do not need to be induced in the rotor (as with, for example, the induction motor), making them somewhat more efficient and giving a slightly greater specific power. However, the permanent magnet rotor does add significantly to the cost of these motors.

6.3.3 Switched reluctance motors

Although only recently coming into widespread use, the switched reluctance (SR) motor is, in principle, quite simple. The basic operation is shown in Figure 6.27. In Figure 6.27(a) the iron stator and rotor are magnetised by a current through the coil on the stator. Because the rotor is out of line with the magnetic field a torque will be produced to minimise the

Figure 6.26 100 kW, oil cooled BLDC motor for automotive application. This unit weighs just 21 kg (photograph reproduced by kind permission of Zytek Ltd.)

air gap and make the magnetic field symmetrical. We could lapse into rather medieval science and say that the magnetic field is 'reluctant' to cross the air gap, and seeks to minimise it. Medieval or not, this is why this type of motor is called a reluctance motor.

At the point shown in Figure 6.27(b) the rotor is aligned with the stator, and the current is switched off. Its momentum then carries the rotor on round over one-quarter of a turn, to the position of 6.27(c). Here the magnetic field is re-applied, in the same direction as before. Again, the field exerts a torque to reduce the air gap and make the field symmetrical, which pulls the rotor on round. When the rotor lines up with the stator again, the current would be switched off.

In the switched reluctance motor, the rotor is simply a piece of magnetically soft iron. Also, the current in the coil does not need to alternate. Essentially then, this is a very simple and potentially low cost motor. The speed can be controlled by altering the length of time that the current is on for in each power pulse. Also, since the rotor is not a permanent magnet, there is no back EMF. generated in the way it is with the BLDC motor, which means that higher speeds are possible. In the fuel cell context, this makes the SR motor particularly suitable for radial compressors and blowers.

The main difficulty with SR motor is that the timing of the turning on and off of the stator currents must be much more carefully controlled. For example, if the rotor is 90° out of line, as in Figure 6.24(a), and the coil is magnetised, no torque will be produced, as the field would be symmetrical. So, the torque is much more variable, and as a result early SR motors had a reputation for being noisy.

The torque can be made much smoother by adding more coils to the stator. The rotor is again laminated iron, but has 'salient poles', i.e. protruding lumps. The number of salient poles will often be two less than the number of coils. Figure 6.28 shows the principle. In 6.28(a) coil A is magnetised, exerting a clockwise force on the rotor. When the salient poles are coming into line with coil A, the current in A is switched off. Two other

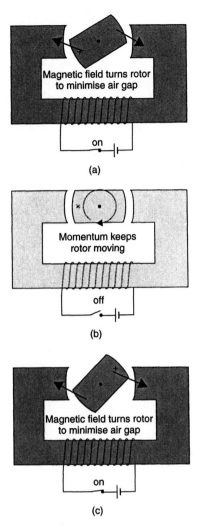

Figure 6.27 Diagram showing the principle of operation of the switched reluctance motor

salient poles are now nearly in line with coil C, which is energised, keeping the rotor smoothly turning. Correct turning on and off of the currents in each coil clearly needs good information about the position of the rotor. This is usually provided by sensors, but modern control systems can do without these. The position of the rotor is inferred from the voltage and current patterns in the coils. This clearly requires some very rapid and complex analysis of the voltage and current waveforms, and is achieved using a special type of microprocessor called a digital signal processor.[5]

---

[5] Although they were originally conceived as devices for processing audio and picture signals, the control of motors is now a major application of digital signal processors. BLDC motors can also operate without rotor position sensors in a similar way.

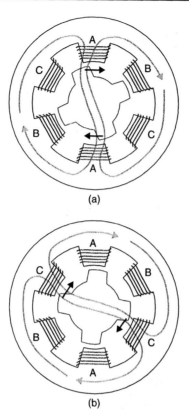

Figure 6.28 Diagram showing the operation of an SR motor with a four salient pole rotor

An example of a rotor and stator from an SR motor is shown in Figure 6.29. In this example the rotor has eight salient poles.

The stator of an SR motor is similar to that in both the induction and BLDC motor. The control electronics are also similar: a microprocessor and some electronic switches, along the lines of Figure 6.21. However, the rotor is significantly simpler, and so cheaper and more rugged. Also, when using a core of high magnetic permeability the torque that can be produced within a given volume exceeds that produced in induction motors (magnetic action on current) and BLDC motors (magnetic action on permanent magnets) (Kenjo 1991, p. 161). Combining this with the possibilities of higher speed means that a higher power density is possible. The greater control precision needed for the currents in the coils makes these motors somewhat harder to apply on a 'few-of' basis, with the result that they are most widely used in cost-sensitive mass-produced goods such as washing machines and food processors. However, we can be sure that their use will become much more widespread.

Although the peak efficiency of the SR motor may be slightly below that of the BLDC motor, SR motors maintain their efficiency over a wider range of speed and torque than any other motor type.

Figure 6.29 The rotor and stator from an SR motor (photograph reproduced by kind permission of SR Drives Ltd.)

6.3.4 The induction motor

The induction motor is very widely used in industrial machines of all types. Its technology is very mature. Induction motors require an AC supply, which might make them seem unsuitable for a DC source such as batteries or fuel cells. However, as we have seen, AC can easily be generated using an inverter, and in fact the inverter needed to produce the AC for an induction motor is no more complicated or expensive than the circuits needed to drive the brushless DC or switched reluctance motors we have just described. So, these widely available and very reliable motors are well suited to use in electric vehicles.

The principle of operation of the three-phase induction motor is shown in Figures 6.30 and 6.31. Three coils are wound right around the outer part of the motor, known as the stator, as shown in the top of Figure 6.30. The rotor usually consists of copper or aluminium rods, all electrically linked (short circuited) at the end, forming a kind of cage, as also shown in Figure 6.30. Although shown hollow, the interior of this cage rotor will usually be filled with laminated iron.

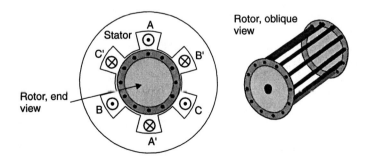

Figure 6.30 Diagram showing the stator and rotor of an induction motor

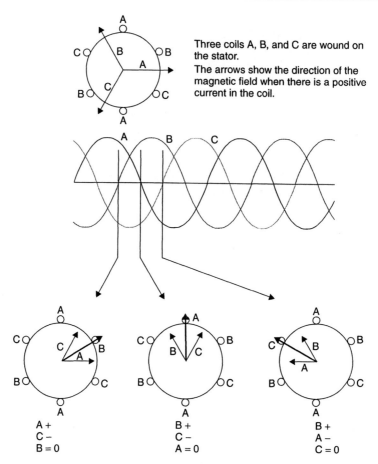

Figure 6.31 Diagrams to show how a rotating magnetic field is produced within an induction motor

The three windings are arranged so that a positive current produces a magnetic field in the direction shown in Figure 6.31. If these three coils are fed with a three-phase alternating current, as in Figure 6.23, the resultant magnetic field rotates anti-clockwise, as shown at the bottom of Figure 6.31.

This rotating field passes through the conductors on the rotor, generating an electric current.

A force is produced on these conductors carrying an electric current, which turns the rotor. It tends to 'chase' the rotating magnetic field. If the rotor were to go at the same speed as the magnetic field, there would be no relative velocity between the rotating field and the conductors, and so no induced current, and no torque. The result is that the torque speed graph for an induction motor has the characteristic shape shown in Figure 6.32. The torque rises as the angular speed slips behind that of the magnetic field, up to an optimum slip, after which the torque declines somewhat.

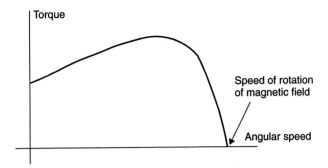

Figure 6.32 Typical torque/speed curve for an induction motor

The winding arrangement of Figure 6.30 and 6.31 is knows as two-pole. It is possible to wind the coils so that the magnetic field has four, six, eight or any even number of poles. The speed of rotation of the magnetic field is the supply frequency divided by the number of pole pairs. So, a four-pole motor will turn at half the speed of a two-pole motor, given the same frequency AC supply, a six-pole motor one-third of the speed, and so on. This gives a rather inflexible way of controlling speed. A much better way is to control the frequency of the three-phase supply. Using a circuit such as that of Figure 6.21, this is easily done. The frequency does not precisely control the speed, as there is a slip, depending on the torque. However, if the angular speed is measured, and incorporated into a feedback loop, the frequency can be adjusted to attain the desired speed.

The maximum torque depends on the strength of the magnetic field in the gap between the rotor and the coils on the stator. This depends on the current in the coils. A problem is that as the frequency increases the current reduces, if the voltage is constant, because of the inductance of the coils having an impedance that is proportional to the frequency. The result is that, if the inverter is fed from a fixed voltage, the maximum torque is inversely proportional to the speed. This is liable to be the case with a fuel cell or battery system.

Induction motors are very widely used. A very high volume of production makes for a very reasonably priced product. Much research has gone into developing the best possible materials. Induction motors are as reliable and well developed as any technology. However, the fact that a current has to be induced in the rotor adds to the losses, with the result that induction motors tend to be a little (1 or 2%) less efficient than the other brushless types, all other things being equal.

6.4 Motor Cooling, Efficiency, Size and Mass

6.4.1 Improving motor efficiency

It is clear that the motor chosen for any application should be as efficient as possible. How can we predict what the efficiency of a motor might be? It might be supposed that the *type* of motor chosen would be a major factor, but in fact it is not. Other factors are much more influential than whether the motor is BLDC, SR or induction.

An electric motor is, in energy terms, fairly simple. Electrical power is the input, and mechanical work is the desired output, with some of the energy being converted into heat. The input and output powers are straightforward to measure, the product of voltage and current for the input, and torque and angular speed at the output. However, the efficiency of an electric motor is not so simple to measure and describe as might be supposed. The problem is that it can change markedly with different conditions, and there is no single internationally agreed method of stating the efficiency of a motor[6] (Auinger 1999). Nevertheless it is possible to state some general points about the efficiency of electric motors, the advantages and disadvantages of the different types, and the effect of motor size. In Section 6.1.5 we also generated a general formula (equation (6.14)) for the efficiency of an electric motor that holds quite well for all motor types.

The first general point is that motors become more efficient as their *size* increases. Table 6.2 shows the efficiency of a range of three-phase, four-pole induction motors. The efficiencies given are the minimum to be attained before the motor can be classified Class 1 efficiency under European Union regulations. The figures clearly show the effect of size. While these figures are for induction motors, exactly the same effect can be seen with other motor types, including BLDC and SR.

The second factor that has more control over efficiency than motor type is the *speed* of a motor. Higher speed motors are more efficient than lower speed ones. The reason for this is that one of the most important losses in a motor is proportional to torque, rather than power, and a lower speed motor will have a higher torque, for the same power, and hence higher losses.

A third important factor is the cooling method. Motors that are liquid cooled run at lower temperatures, which reduces the resistance of the windings, and hence improves efficiency, though this will only affect things by about 1%.

Another important consideration is that the efficiency of an electric motor might well be very different from any figure given in the specification, if it operates well away from optimum speeds and torque. In some cases an efficiency map, like that of Figures 6.7 and 6.33

Table 6.2 The minimum efficiency of four-pole three-phase induction motors to be classified as Class 1 efficiency under EU regulations. Efficiency measured according to IEC 36.2

Power, kW	Minimum efficiency, %
1.1	83.8
2.2	86.4
4	88.3
7.5	90.1
15	91.8
30	93.2
55	96.2
90	95.0

---

[6] The nearest to such a standard is IEC 34-2.

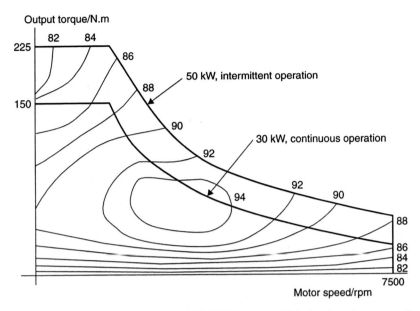

Figure 6.33 The efficiency map for a 30 kW BLDC motor. This is taken from manufacturer's data, but note that in fact at zero speed the efficiency must be 0%

may be provided. That given in Figure 6.33 is based on a real BLDC motor. The maximum efficiency is 94%, but this efficiency is only obtained for a fairly narrow range of conditions. It is quite possible for the motor to operate at well below 90% efficiency.

As a general guide, we can say that the maximum efficiency of a good quality motor will be quite close to the figures given in Table 6.2 for all motor types, even if they are not induction motors. The efficiency of the BLDC and SR motors is likely to be one or two percent higher than for an induction motor, since there is less loss in the rotor. The SR motor manufacturers also claim that their efficiency is maintained over a wider range of speed and torque conditions.

6.4.2 Motor mass

A motor should generally be as small and light as possible, while delivering the required power. As with the case of motor efficiency, the type of motor chosen is much less important than other factors (such as cooling method and speed) when it comes to the specific power and power density of an electric motor. The one exception to this is the brushed DC motor. We explained in Section 6.1.6 that the brushed DC motor is bound to be rather larger that other types, because such a high proportion of the losses are generated in the rotor, in the middle of the motor.

Figure 6.34 is a chart showing typical specific powers for different types of motor at different powers. Taking the example of the BLDC motor, it can be seen that the cooling method used is a very important factor. The difference between the air cooled and liquid cooled BLDC motor is most marked. The reason for this is that the motor has to be large enough to dispose of the heat losses. If the motor is liquid cooled, then the same heat losses can be removed from a smaller motor.

We would then expect that efficiency should be an important factor. A more efficient motor could be smaller, since less heat disposal would be needed. This is indeed the case, and as a result all the factors that produce higher efficiency, and which were discussed in the previous section, also lead to greater specific power. The most important of these are as follows.

Higher *power* leads to higher efficiency, and hence higher specific power. This can be very clearly seen in Figure 6.34. (However, note that the logarithmic scale tends to make this effect appear less marked.)

Higher *speed* leads to higher power density. The size of the motor is most strongly influenced by the motor *torque* than *power*. The consequence is that a higher speed, lower torque motor will be smaller. So if a low speed rotation is needed, a high speed motor with a gearbox will be lighter and smaller than a low speed motor. A good example is an electric vehicle, where it would be possible to use a motor directly coupled to the axle. However, this is not often done, and a higher speed motor is connected by (typically) a 10:1 gearbox. Table 6.3 shows this, by giving the mass of a sample of induction motors of the same power but different speeds.

The more efficient *motor types*, SR and BLDC, have higher power density that the induction motor.

The curves of Figure 6.34 give a good idea of the likely power density that can be expected from a motor, and can be used to estimate the mass. The lines are necessarily

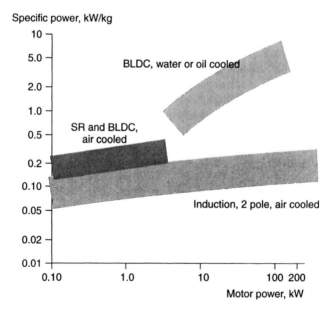

Figure 6.34 Chart to show the specific power of different types of electric motor at different powers. The power here is the continuous power. Peak specific powers will be about 50% higher. Note the logarithmic scales (this chart was made using data from several motor manufacturers.)

Table 6.3 The mass of some 37 kW induction motors, from the same manufacturer, for different speeds. The speed is for a 50 Hz AC supply

Speed (rpm)	Mass (kg)
3000	270
1500	310
1000	415
750	570

broad, as the mass of a motor will depend on many factors other than those we have already discussed. The material the frame is made from is of course very important, as is the frame structure.

6.5 Electrical Machines for Hybrid Vehicles

The motors and alternators used in hybrid electric vehicles are in principle no different from those described above. Indeed in many cases there is no significant difference between the motors used in hybrid vehicles than any other type.

The basic principles of some types of hybrid vehicle were described in Chapter 1. In the series hybrid vehicle there is really nothing different about the electrical machines from those used in a host of other applications. The traction motor, for example, will work in the same way as in the case of the classic battery powered electric vehicle.

It is in the parallel hybrid that there is scope for some novelty in machine design. One example is the crankshaft mounted electrical machine that is used in a number of designs, including the groundbreaking Honda Insight. Here the electrical machine, which can work as either a motor or generator, is mounted directly in line with the engine crankcase. Such machines are in most cases a type of brushless DC (or synchronous AC) motor, as described in Section 6.3.2. They will be multiple-pole machines, since their location means their dimensions need to be short in length and wide in diameter. They are nearly always different from the machine of Figure 6.25 in one important respect: they are usually 'turned inside out', with the stationary coils being on the inside, and the rotor being a band of magnets moving outside the coil. The idea is shown in Figure 6.35. The larger diameter permits this construction, which has the advantage that the centrifugal force on the magnets tends to make them stay in place, rather than throw them out of their mounting.

It is worth pointing out that this same type of inside out motor is used in motors that are integral with wheels, such as the machine of Figure 8.9.

However, not all parallel hybrids use special multiple motors of this type. Some hybrid vehicles the electrical machine use a fairly conventional, single pole, fairly high speed machine, which is connected to the engine crankshaft much like the alternator in an ordinary conventional IC engine vehicle. The fan belt type connection is made rather more robust.

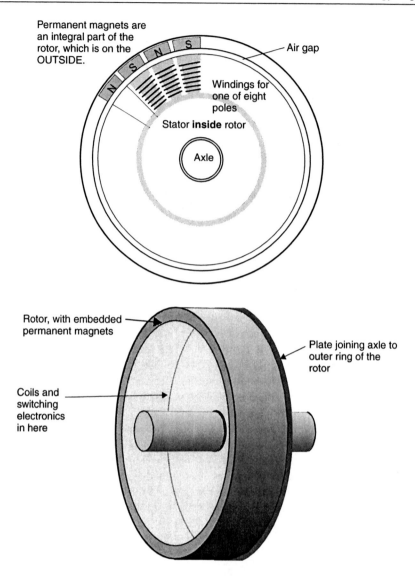

Figure 6.35 Diagram of inside out electric motor

Another type of parallel hybrid where fairly conventional motors are used is the type where the front wheels are driven by an electric motor, and the rear are driven directly from the IC engine. The front wheels are electrically powered when more power is needed, or when driving very slowly in a queue, or when four wheel drive is required for traction purposes. Similarly, braking is provided by the front axle, using the machine as a generator, regenerating some of the energy when slowing down. This type of parallel hybrid may well be suitable for some larger cars and vans, and is illustrated in Figures 1.12 and 1.13.

Electric Machines and their Controllers

Figure 6.36 Demonstration hybrid diesel/electric power unit installed in a Smart. The motor/generator is the unit marked 'hyper' at the base. (Photograph reproduced by kind permission of Micro Compact Car smart GmbH.)

The machine driving, and being driven by, the front axle need not be any different from those described in Section 6.3.

Yet another parallel hybrid arrangement that has been tried with some success on the Smart car is shown in Figure 6.36. Here the main engine was left almost unaltered, but a motor was added near the base of the engine, so that it connects to the drive differential via its own drive system. This allows it to drive the vehicle if the engine is not working, yet it does not have to be in line with the crankshaft. Minimal alterations are made to the rest of the engine, which reduces cost. The shape and working speeds of the electric motor mean that it does not need to be of a very special type.

References

Auinger H. (1999) Determination and designation of the efficiency of electrical machines. *Power Engineering Journal*, Vol. 13, No. 1, pp. 15–23.
Kenjo T. (1991) *Electric Motors and their Controls*, Oxford University Press.
Kenjo T. (1994) *Power Electronics for the Microprocessor Age*, Oxford University Press.
Spiegel R.J., Gilchrist T. and House D.E. (1999) Fuel cell bus operation at high altitude. *Proceedings of the Institution of Mechanical Engineers*, Vol. 213. part A, pp. 57–68.
Walters D. (1999) Energy efficient motors, saving money or costing the earth?. *Power Engineering Journal*, Part 1, Vol. 13, No. 1, pp. 25–30, Part 2, Vol. 13, No. 2, pp. 44–48.

7

Electric Vehicle Modelling

7.1 Introduction

With all vehicles the prediction of performance and range is important. Computers allow us to do this reasonably easily. Above all, computer based methods allow us to quickly experiment with aspects of the vehicle, such as motor power, battery type and size, weight and so on, and see how the changes affect the performance and range. In this chapter we will show how the equations we have developed in the preceding chapters can be put together to perform quite accurate and useful simulations. Furthermore, we will show how this can be done without using any special knowledge of programming techniques, as standard mathematics and spreadsheet programs such as MATLAB® and EXCEL® make an excellent basis for these simulations. We will also see that there are some features of electric vehicles that make the mathematical modelling of performance easier than for other vehicles.

The first parameter we will model is vehicle *performance*. By performance we mean acceleration and top speed, an area where electric vehicles have a reputation of being very poor. It is necessary that any electric vehicle has a performance that allows it, at the very least, to blend safely with ordinary city traffic. Many would argue that the performance should be at least as good as current IC engine vehicles if large scale sales are to be achieved.

Another vitally important feature of electric vehicles that we must be able to predict is their *range*. This can also be mathematically modelled, and computer programs make this quite straightforward. The mathematics we will develop will allow us to see the effects of changing things like battery type and capacity, as well as all other aspects of vehicle design, on range. This is an essential tool for the vehicle designer.

We will go on to show how the data produced by the simulations can also have other uses in addition to predicting performance and range. For example we will see how data about the motor torque and speed can be used to optimise the compromises involved in the design of the motor and other subsystems.

Electric Vehicle Technology Explained   James Larminie and John Lowry
© 2003 John Wiley & Sons, Ltd   ISBN: 0-470-85163-5

7.2 Tractive Effort

7.2.1 Introduction

The first step in vehicle performance modelling is to produce an equation for the tractive effort. This is the force propelling the vehicle forward, transmitted to the ground through the drive wheels.

Consider a vehicle of mass m, proceeding at a velocity v, up a slope of angle ψ, as in Figure 7.1. The force propelling the vehicle forward, the tractive effort, has to accomplish the following:

- overcome the rolling resistance;
- overcome the aerodynamic drag;
- provide the force needed to overcome the component of the vehicle's weight acting down the slope;
- accelerate the vehicle, if the velocity is not constant.

We will consider each of these in turn.

7.2.2 Rolling resistance force

The rolling resistance is primarily due to the friction of the vehicle tyre on the road. Friction in bearings and the gearing system also play their part. The rolling resistance is approximately constant, and hardly depends on vehicle speed. It is proportional to vehicle weight. The equation is:

$$F_{rr} = \mu_{rr} mg \tag{7.1}$$

where μ_{rr} is the coefficient of rolling resistance. The main factors controlling μ_{rr} are the type of tyre and the tyre pressure. Any cyclist will know this very well; the free-wheeling performance of a bicycle becomes much better if the tyres are pumped up to a high pressure, though the ride may be less comfortable.

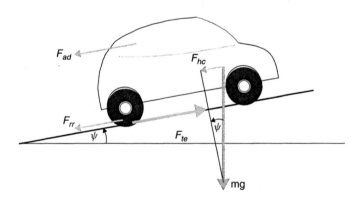

Figure 7.1 The forces acting on a vehicle moving along a slope

The value of μ_{rr} can reasonably readily be found by pulling a vehicle at a steady very low speed, and measuring the force required.

Typical values of μ_{rr} are 0.015 for a radial ply tyre, down to about 0.005 for tyres developed especially for electric vehicles.

7.2.3 Aerodynamic drag

This part of the force is due to the friction of the vehicle body moving through the air. It is a function of the frontal area, shape, protrusions such as side mirrors, ducts and air passages, spoilers, and many other factors. The formula for this component is:

$$F_{ad} = \tfrac{1}{2}\rho A C_d v^2 \tag{7.2}$$

where ρ is the density of the air, A is the frontal area, and v is the velocity. C_d is a constant called the drag coefficient.

The drag coefficient C_d can be reduced by good vehicle design. A typical value for a saloon car is 0.3, but some electric vehicle designs have achieved values as low as 0.19. There is greater opportunity for reducing C_d in electric vehicle design because there is more flexibility in the location of the major components, and there is less need for cooling air ducting and under-vehicle pipework. However, some vehicles, such as motorcycles and buses will inevitably have much larger values, and C_d figures of around 0.7 are more typical in such cases.

The density of air does of course vary with temperature, altitude and humidity. However a value of $1.25 \,\text{kg.m}^{-3}$ is a reasonable value to use in most cases. Provided that SI units are used (m^2 for A, m.s^{-1} for v) then the value of F_{ad} will be given in Newtons.

7.2.4 Hill climbing force

The force needed to drive the vehicle up a slope is the most straightforward to find. It is simply the component of the vehicle weight that acts along the slope. By simple resolution of forces we see that:

$$F_{hc} = mg \sin(\psi) \tag{7.3}$$

7.2.5 Acceleration force

If the velocity of the vehicle is changing, then clearly a force will need to be applied in addition to the forces shown in Figure 7.1. This force will provide the *linear acceleration* of the vehicle, and is given by the well-known equation derived from Newton's second law,

$$F_{la} = ma \tag{7.4}$$

However, for a more accurate picture of the force needed to accelerate the vehicle we should also consider the force needed to make the rotating parts turn faster. In other words, we need to consider *rotational* acceleration as well as *linear* acceleration. The main issue here is the electric motor, not necessarily because of its particularly high moment of inertia, but because of its higher angular speeds.

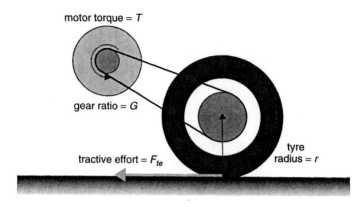

Figure 7.2 A simple arrangement for connecting a motor to a drive wheel

Referring to Figure 7.2, clearly the axle torque $= F_{te}r$, where r is the radius of the tyre, and F_{te} is the tractive effort delivered by the powertrain. If G is the gear ratio of the system connecting the motor to the axle, and T is the motor torque, then we can say that:

$$T = \frac{F_{te}r}{G}$$

and $\quad F_{te} = \frac{G}{r}T \quad$ (7.5)

We will use this equation again when we develop final equations for vehicle performance. We should also note that:

$$\text{axle angular speed} = \frac{v}{r} \text{ rad.s}^{-1}$$

So motor angular speed

$$\omega = G\frac{v}{r} \text{ rad.s}^{-1} \quad (7.6)$$

Similarly, motor angular acceleration

$$\dot{\omega} = G\frac{a}{r} \text{ rad.s}^{-2}$$

The torque required for this angular acceleration is:

$$T = IG\frac{a}{r}$$

where I is the moment of inertia of the rotor of the motor. The force at the wheels needed to provide the angular acceleration ($F_{\omega a}$) is found by combining this equation

Electric Vehicle Modelling

with equation (7.5), giving:

$$F_{\omega a} = \frac{G}{r} I G \frac{a}{r}$$

$$F_{\omega a} = I \frac{G^2}{r^2} a \qquad (7.7)$$

We must note that in these simple equations we have assumed that the gear system is 100% efficient, it causes no losses. Since the system will usually be very simple, the efficiency is often very high. However, it will never be 100%, and so we should refine the equation by incorporating the gear system efficiency η_g. The force required will be slightly larger, so equation (7.7) can be refined to:

$$F_{\omega a} = I \frac{G^2}{\eta_g r^2} a \qquad (7.8)$$

Typical values for the constants here are 40 for G/r and 0.025 kg.m² for the moment of inertia. These are for a 30 kW motor, driving a car which reaches 60 kph at a motor speed of 7000 rpm. Such a car would probably weigh about 800 kg. The IG^2/r^2 term in equation (7.8) will have a value of about 40 kg in this case. In other words the angular acceleration force given by equation (7.8) will typically be much smaller than the linear acceleration force given by equation (7.4). In this specific (but reasonably typical) case, it will be smaller by the ratio:

$$\frac{40}{800} = 0.05 = 5\%$$

It will quite often turn out that the moment of inertia of the motor I will not be known. In such cases a reasonable approximation is to simply increase the mass by 5% in equation (7.4), and to ignore the $F_{\omega a}$ term.

7.2.6 Total tractive effort

The total tractive effort is the sum of all these forces:

$$F_{te} = F_{rr} + F_{ad} + F_{hc} + F_{la} + F_{\omega a} \qquad (7.9)$$

where:

- F_{rr} is the rolling resistance force, given by equation (7.1);
- F_{ad} is the aerodynamic drag, given by equation (7.2);
- F_{hc} is the hill climbing force, given by equation (7.3);
- F_{la} is the force required to give linear acceleration given by equation (7.4);
- $F_{\omega a}$ is the force required to give angular acceleration to the rotating motor, given by equation (7.8).

We should note that F_{la} and $F_{\omega a}$ will be negative if the vehicle is slowing down, and that F_{hc} will be negative if it is going downhill.

7.3 Modelling Vehicle Acceleration

7.3.1 Acceleration performance parameters

The acceleration of a car or motorcycle is a key performance indicator, though there is no standard measure used. Typically the time to accelerate from standstill to 60 mph, or 30 or 50 kph will be given. The nearest to such a standard for electric vehicles are the 0–30 kph and 0–50 kph times, though these times are not given for all vehicles.

Such acceleration figures are found from simulation or testing of real vehicles. For IC engined vehicles this is done at maximum power, or 'wide open throttle' (WOT). Similarly, for electric vehicles performance simulations are carried out at maximum torque.

We have already seen in Chapter 6 that the maximum torque of an electric motor is a fairly simple function of angular speed. In most cases, at low speeds, the maximum torque is a constant, until the motor speed reaches a critical value ω_c after which the torque falls. In the case of a brushed shunt or permanent magnet DC motor the torque falls linearly with increasing speed. In the case of most other types of motor, the torque falls in such a way that the power remains constant.

The angular velocity of the motor depends on the gear ratio G and the radius of the drive wheel r as in equation (7.6) derived above. So we can say that:

$$\text{For } \omega < \omega_c, \text{ or } v < \frac{r}{G}\omega_c \text{ then } T = T_{\max}$$

Once this constant torque phase is passed, i.e. $\omega \geq \omega_c$, or $v \geq r\omega_c/G$, then *either* the power is constant, as in most brushless type motors, and we have:

$$T = \frac{T_{\max}\omega_c}{\omega} = \frac{rT_{\max}\omega_c}{Gv} \tag{7.10}$$

or the torque falls according to the linear equation we met in Section 6.1.2:

$$T = T_0 - k\omega$$

which, when equation (7.6) is substituted for angular speed, gives

$$T = T_0 - \frac{kG}{r}v \tag{7.11}$$

Now that we have the equations we need, we can combine them in order to find the acceleration of a vehicle. Many of these equations may look quite complex, but nearly all the terms are constants, which can be found or estimated from vehicle or component data.

For a vehicle on level ground, with air density 1.25 kg.m^{-3}, equation (7.9) becomes:

$$F_{te} = \mu_{rr}mg + 0.625AC_dv^2 + ma + I\frac{G^2}{\eta_g r^2}a$$

Electric Vehicle Modelling

Substituting equation (7.5) for F_{te}, and noting that $a = dv/dt$, we have:

$$\frac{G}{r}T = \mu_{rr}mg + 0.625AC_dv^2 + \left(m + I\frac{G^2}{\eta_g r^2}\right)\frac{dv}{dt} \quad (7.12)$$

We have already noted that T, the motor torque, is either a constant of a simple function of speed (equations (7.10) and (7.11)). So, equation (7.13) can be reduced to a differential equation, first order, for the velocity v. Thus the value of v can be found for any value of t.

For example, in the initial acceleration phase, when $T = T_{\max}$, equation (7.12) becomes:

$$\frac{G}{r}T_{\max} = \mu_{rr}mg + 0.625AC_dv^2 + \left(m + I\frac{G^2}{\eta_g r^2}\right)\frac{dv}{dt} \quad (7.13)$$

Provided that all the constants are known, or can reasonably be estimated, this is a very straightforward first-order differential equation, whose solution can be found using many modern calculators, as well as a wide range of personal computer programs. This is also possible for the situation with the larger motors. Two examples will hopefully make this clear.

7.3.2 Modelling the acceleration of an electric scooter

For our first example we will take an electric scooter. No particular model is being taken, but the vehicle is similar to the electric scooters made by Peugeot and EVS, an example of which is shown in Figure 7.3.

- The electric scooter has a mass of 115 kg, with a typical passenger of mass 70 kg, so total mass $m = 185$ kg.
- The moment of inertia of the motor is not known, so we will adopt the expedient suggested at the end of Section 7.2.5, and increase m by 5% in the linear acceleration term only. A value of 194 kg will thus be used from m in the final term of equation (7.13).
- The drag coefficient C_d is estimated as 0.75, a reasonable value for a small scooter, with a fairly 'sit-up' riding style.
- The frontal area of vehicle and rider $= 0.6 \text{ m}^2$.
- The tyres and wheel bearings give a coefficient of rolling resistance, $\mu_{rr} = 0.007$.
- The motor is connected to the rear wheel using a 2:1 ratio belt system, and the wheel diameter is 42 cm. Thus $G = 2$ and $r = 0.21$ m.
- The motor is an 18 V Lynch type motor, of the type discussed in Section 6.1.2. Equation (6.8) has been recalculated for 18 V, giving:

$$T = 153 - 1.16\omega \quad (7.14)$$

- As in Section 6.1.2, the maximum current is controlled by the maximum safe current, in this case 250 A, so, as shown in equation (6.9), the maximum torque $T_{\max}$ is 34 Nm.

Figure 7.3 Electric scooter of the type simulated at various points in this chapter. The photograph was taken in a Berlin car park

- The critical motor speed ω_c after which the torque falls according to equation (7.14) occurs when:

$$34 = 153 - 1.16\omega$$
$$\therefore \omega = \frac{153 - 34}{1.16} = 103 \, \text{rad.s}^{-1}$$

- The gear system is very simple, and of low ratio, and so we can assume a good efficiency. A value of η_g of 0.98 is estimated. An effect of this will be to reduce the torque, and so this factor will be applied to the torque.

When the torque is constant, equation (7.13) becomes:-

$$\frac{2}{0.21} \times 0.98 \times 34 = 0.007 \times 185 \times 9.8 + 0.625 \times 0.6 \times 0.75v^2 + 194\frac{dv}{dt}$$
$$317 = 12.7 + 0.281v^2 + 194\frac{dv}{dt}$$

thus $194\dfrac{dv}{dt} = 304 - 0.281v^2$

so $\dfrac{dv}{dt} = 1.57 - 0.00145v^2$ \hfill (7.15)

This equation holds until the torque begins to fall when, $\omega = \omega_c = 103\,\text{rad s}^{-1}$, which corresponds to $103 \times 0.21/2 = 10.8\,\text{ms}^{-1}$. After this point the torque is governed by equation (7.14). If we substitute this, and the other constants, into equation (7.12) we obtain:

$$\dfrac{2}{0.21} \times 0.98 \times \left(153 - 1.16\dfrac{2}{0.21}v\right) = 0.007 \times 185 \times 9.8 + 0.625 \times 0.6 \times 0.75v^2 + 194\dfrac{dv}{dt}$$

$$\therefore 1428 - 103v = 12.7 + 0.281v^2 + 194\dfrac{dv}{dt}$$

and so $\dfrac{dv}{dt} = 7.30 - 0.53v - 0.00145v^2$ \hfill (7.16)

There are many practical and simple ways of solving these differential equations. Many modern calculators will solve such equations, remembering that there is a simple initial condition that $v = 0$ when $t = 0$. However, the most versatile next step is to derive a simple numerical solution, which can then easily be used in EXCEL® or MATLAB®. The derivative of v is simply the difference between consecutive values of v divided by the time step. Applying this to equation (7.15) gives us:

$$\dfrac{v_{n+1} - v_n}{\partial t} = 1.57 - 0.00145v_n^2$$

For a program such as EXCEL® or MATLAB® we need to rearrange this equation to obtain the value of the next velocity from the current velocity. This is done as follows:

$$v_{n+1} = v_v + \partial t \times (1.57 - 0.00145v_n^2) \hfill (7.17)$$

This equation holds for velocities up to the critical velocity of $10.8\,\text{ms}^{-1}$, after which we have to use equation (7.16), approximated in exactly the same way as we have just done for equation (7.15), which gives:

$$v_{n+1} = v_v + \partial t \times (7.30 - 0.53v - 0.00145v_n^2) \hfill (7.18)$$

The MATLAB® script file below shows how to solve these equations using this program. Figure 7.4 is a plot of the solution using a time step ∂t of 0.1 s. Exactly the same result can be obtained with almost equal ease using EXCEL®. It is left as an exercise for the reader to produce an EXCEL® spreadsheet or MATLAB® script file where many of the machine parameters, such as the gear ratio G, are left as easily altered variables, so that the effect of changing them on the vehicle's performance can be noted.

```
% ScootA - electric scooter acceleration.
t=linspace(0,50,501);   % 0 to 50 s, in 0.1 s steps
vel=zeros(1,501);       % 501 readings of velocity
d=zeros(1,501);%Array for storing distance traveled
dT=0.1;         % 0.1 second time step

for n= 1:500
   % Now follow equations 7.17 & 7.18
   if vel(n)<10.8 % Torque constant till this point
      vel(n+1) = vel(n) + dT*(1.57 - (0.00145*(vel(n)^2)));
   elseif vel(n)>=10.8
        vel(n+1)=vel(n)+dT*(7.30-(0.53*vel(n))-
        (0.00145*(vel(n)^2)));
   end;
d(n+1)=d(n) + 0.1*vel(n); % Compute distance traveled.
end;
vel=vel.*3.6; % Multiply by 3.6 to convert m/sec to kph
plot(t,vel); axis([0 30 0 50]);
xlabel('Time/seconds');
ylabel('Velocity/kph');
title('Full power (WOT) acceleration of electric scooter');
```

The result of this simulation is shown in Figure 7.4, and this shows that the performance is somewhat as might be expected from a fairly low power motor. The acceleration is unspectacular, and the top speed is about 30 mph, or 48 kph, on level ground. However, this is reasonably compatible with safe city riding. The acceleration of such vehicles is sometimes given in terms of the standing start 100 m times, and the power of such

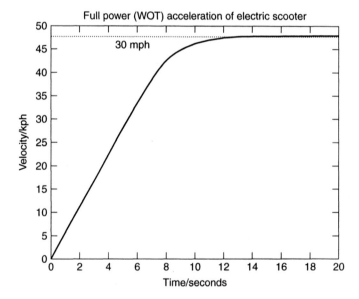

Figure 7.4 A graph showing the acceleration of a design of electric scooter, being the solution of equations (7.15) and (7.16), as approximated by equations (7.17) and (7.18), with a 0.1 s time step

Electric Vehicle Modelling

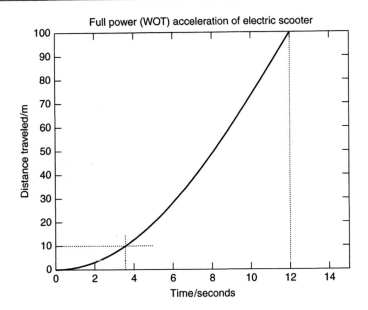

Figure 7.5 Distance/time graph for an electric scooter, showing the time to cover 10 m and 100 m from a standing start

MATLAB® script files is that they can very easily be changed to produce such information. If the plot line in the file above is changed as follows, then Figure 7.5 is obtained.

```
plot(t,d); axis([0 15 0 100]);
```

While we are not claiming that our model exactly represents any particular commercial designs, it is worth noting the following points from the specification of the Peugeot 'SCOOT'ELEC' performance specification:

- maximum speed 45 kph, 28 mph;
- 10 m from standing start time, 3.2 s
- 100 m from standing start time, 12 s

It is very clear from Figures 7.4 and 7.5 that the performance of our simulated vehicle is remarkably similar.

7.3.3 Modelling the acceleration of a small car

For our second example we will use a vehicle that had an important impact on the recent development of electric cars. The GM EV1 was arguably the first modern electric car from one of the really large motor companies. It incorporated technologies that were quite novel for its time, and is indeed still unsurpassed as a design of battery electric car. Several views of this vehicle are shown in Figure 11.5. Further details of this car are given in Section 11.2, but as far as simulating its performance, the main features are:

- an ultra-low drag coefficient C_d of 0.19;
- a very low coefficient of rolling resistance μ_{rr} of 0.0048;
- the use of variable frequency induction motors, operating at very high speed, nearly 12 000 rpm at maximum speed.

Further data is taken from company information[1] about the vehicle:

- vehicle mass = 1400 kg.[2] Then add a driver and a passenger, each weighing 70 kg, giving $m = 1540$ kg;
- the motor's moment of inertia is not known, however, compared to the mass of such a heavy vehicle this will be very low; the wheels are also very light and we will approximate this term by increasing the mass very modestly to 1560 kg in the final term of equation (7.12);
- the gear ratio is 11:1, thus $G = 11$; the tyre radius is 0.30 m;
- for the motor, $T_{\max} = 140$ Nm and $\omega_c = 733$ rad s^{-1} note this means $T = T_{\max}$ until $v = 19.8$ ms^{-1} ($= 71.3$ kph);
- above 19.8 ms^{-1} the motor operates at a constant 102 kW, as this is a WOT test, so:

$$T = \frac{102\,000}{37 \times v} = \frac{2756}{v}$$

- the frontal area $A = 1.8$ m^2;
- the efficiency of the single-speed drive coupling between motor and axle is estimated as 95%, so $\eta_g = 0.95$; the values of the torque T will be reduced by a factor of 0.95; this slightly lower figure is because there is a differential and a higher ratio gear box than in the last example.

These values can now be put into equation (7.12), giving for the first phase while the motor torque is constant:

$$0.95 \times 37 \times 140 = 72.4 + 0.214v^2 + 1560\frac{dv}{dt}$$

$$\text{So } \frac{dv}{dt} = 3.11 - 0.000137v^2 \tag{7.19}$$

Once the speed has reached 19.8 ms^{-1} the velocity is given by the differential equation:

$$0.95 \times 37 \times \frac{2756}{v} = 72.4 + 0.214v^2 + 1560\frac{dv}{dt}$$

$$\text{So } \frac{dv}{dt} = \frac{62.1}{v} - 0.046 - 0.000137v^2 \tag{7.20}$$

---

[1] The two sources are Shnayerson (1996) and the official GM EV1 website at www.gmev.com.
[2] It is interesting to note that 594 kg, or 42%, of this is the lead acid batteries!

Electric Vehicle Modelling

The procedure for finding the acceleration is very similar to the first example; the only extra complication is that when the velocity reaches $35.8\,\text{ms}^{-1}$ it stops rising, as at this point the motor controller limits any further acceleration.

Before any program such as EXCEL® or MATLAB® can be used, the key equations (7.19) and (7.20) must be put into 'finite difference' form. This is done exactly as we did for equations (7.15) and (7.16). The two equations become:

$$v_{n+1} = v_n + \partial t (3.11 - 0.000137 v_n^2) \tag{7.21}$$

$$v_{n+1} = v_n + \partial t \left(\frac{62.1}{v_n} - 0.046 - 0.000137 v_n^2 \right) \tag{7.22}$$

The MATLAB® script file for these equations is very similar to that for the electric scooter given above, so it is not given here in the main text, but can be found in Appendix 1. The plot of velocity against time is shown in Figure 7.6. Looking at Figure 7.6 we can see that the time taken to reach 96 kph, which is 60 mph, is just under 9 s. Not only is this a very respectable performance, but it is also exactly the same as given in the official figures for the performance of the real vehicle.

We have thus seen that, although not overly complex, this method of modelling vehicle performance gives results that are validated by real data. We can thus have confidence in this method. However, vehicles are required to do more than just accelerate well

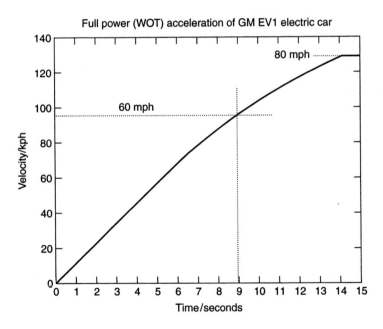

Figure 7.6 A graph of velocity against time for a GM EV1 at full power. This performance graph, obtained from a simple mathematical model, gives very good agreement with published real performance data

from a standing start, and in the next section we tackle the more complex issue of range modelling.

7.4 Modelling Electric Vehicle Range

7.4.1 Driving cycles

It is well known that the range of electric vehicles is a major problem. In the main this is because it is so hard to efficiently store electrical energy. In any case, this problem is certainly a critical issue in the design of any electric vehicle. There are two types of calculation or test that can be performed with regard to the range of a vehicle.

The first, and much the simplest, is the constant velocity test. Of course no vehicle is really driven at constant velocity, especially not on level ground, and in still air, which are almost universal further simplifications for these tests. However, at least the rules for the test are clear and unambiguous, even if the test is unrealistic. It can be argued that they do at least give useful comparative figures.

The second type of test, more useful and complex, is where the vehicle is driven, in reality or in simulation, through a profile of ever changing speeds. These test cycles have been developed with some care, and there are (unfortunately) a large number of them. The cycles are intended to correspond to realistic driving patterns in different conditions. During these tests the vehicle speed is almost constantly changing, and thus the performance of all the other parts of the system is also highly variable, which makes the computations more complex. However, modern computer programs make even these more complex situations reasonably straightforward.

These driving cycles (or schedules) have primarily been developed in order to provide a realistic and practical test for the emissions of vehicles. One of the most well-known of the early cycles was one based on actual traffic flows in Los Angeles CA, and is known as the LA-4 cycle. This was then developed into the Federal Urban Driving Schedule, or FUDS. This is a cycle lasting 1500 seconds, and for each second there is a different speed, as shown in Figure 7.7. There is also a simplified version of this cycle known as SFUDS, shown in Figure 7.8, which has the advantage that it only lasts 360 seconds, and so has only 360 data points. This has the same average speed, the same proportion of time stationary, the same maximum acceleration and braking, and gives very similar results when used for simulating vehicle range.

These cycles simulate urban driving, but other cycles are used to simulate out-of-town or highway driving. Two notable examples of these are the FHDS, shown in Figure 7.9. Although widely used, this cycle has a rather unrealistic maximum speed for highway driving, and the newer US06 standard is now becoming more widely used instead.

In the European scene the cycles tend to be rather simpler, with periods of constant acceleration and constant velocity. Of particular note is the ECE-15 drive cycle, shown in Figure 7.10, which is useful for testing the performance of small vehicles such as battery electric cars. In EC emission tests this has to be combined the extra-urban driving cycles (EUDC), which has a maximum speed of $120 \, \text{km.h}^{-1}$.

Currently the most widely used standard in Asia is the Japanese 10–15 Mode cycle. Like the European cycles, this involves periods of constant velocity and acceleration. It

Electric Vehicle Modelling

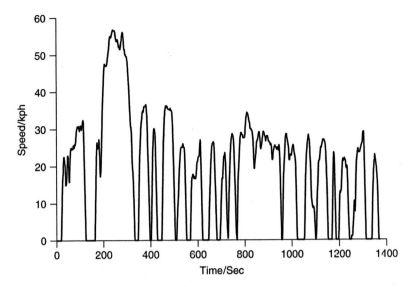

Figure 7.7 The Federal Urban Driving Schedule, as used for emission testing by the United States Environmental Protection Agency

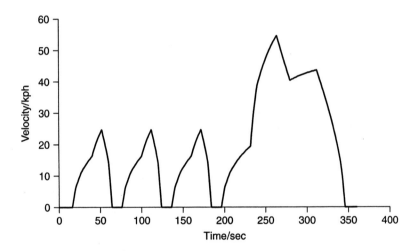

Figure 7.8 Graph of speed against time for the simplified federal urban driving schedule

is not unlike a combination of the European ECE-15 urban driving cycle and the EUDC. At the time of writing, this cycle must be used in stating ranges for vehicles in Japan, as well as for emission tests.

All these standards have maximum speeds in the region of $100\,km.h^{-1}$. For several important types of electric vehicle, including the electric delivery vehicle and the electric

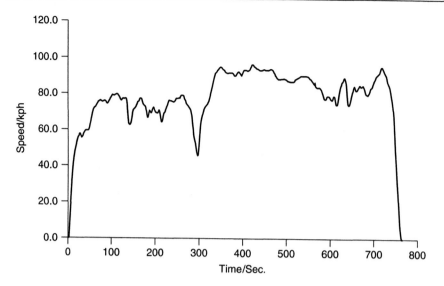

Figure 7.9 Graph of speed against time for the 765 second federal highway driving schedule

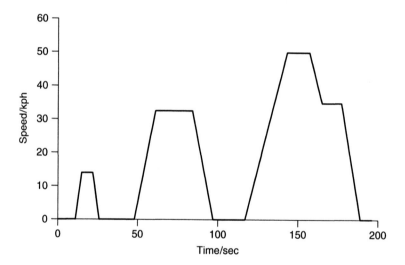

Figure 7.10 European urban driving schedule ECE-15

motor scooter, this is an unrealistic speed, which can often not be achieved. To simulate these vehicles other standard cycles are needed. A fairly old standard, which was developed specifically for electric vehicles in the 1970s, is the SAE J227a driving schedule. This has four versions, with different speeds. Each cycle is quite short in time, and consists of an acceleration phase, a constant velocity phase, a 'coast' phase, and a braking phase, followed by a stationary time. The coasting phase, where the speed is not specified,

Electric Vehicle Modelling

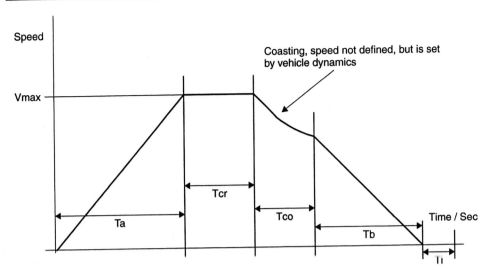

Figure 7.11 Diagram for SAE J227a cycle. The figures for the various times are given in Table 7.1

Table 7.1 Nominal parameters for the four variations of the SAE J227a test schedule. These figures should be read in conjunction with Figure 7.10

Parameter	Unit	Cycle A	Cycle B	Cycle C	Cycle D
Maximum speed	Km.h^{-1}	16	32	48	72
Acceleration Time Ta	s	4	19	18	28
Cruise time Tcr	s	0	19	20	50
Coast time Tco	s	2	4	8	10
Brake time Tb	s	3	5	9	9
Idle time Ti	s	30	25	25	25
Total time	s	39	72	80	122

but the tractive effort is set to zero, is somewhat of a nuisance to model.[3] The general velocity profile is shown in Figure 7.11, and the details of each of the four variants of this cycle are given in Table 7.1 The most commonly used cycle is SAE J227a-C, which is particularly suitable for electric scooters and smaller city-only electric vehicles. The A and B variants are sometimes used for special purpose delivery vehicles.

Another schedule worthy of note for low speed vehicles is the European ECE-47 cycle, which is used for the emission testing of mopeds and motorcycles with engine capacity less than 50 cm^3. It is also widely used for the range simulation of electric scooters. Like the SAEJ227 cycle it can be a little complicated to run the simulation, as the speed is not specified at all times. Instead the vehicle is run from standstill at WOT for 50 s. The vehicle is then slowed to 20 km.h^{-1} over the next 15 s, after which this

---

[3] It is not uncommon to get around this difficulty by simply putting in likely figures for a somewhat gentle period of deceleration.

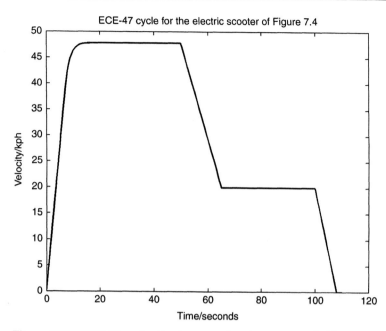

Figure 7.12 ECE-47 cycles for the same electric scooter as in Figure 7.4

speed is maintained for 35 seconds. Finally the vehicle is brought to a halt, at constant deceleration, over the next 8 s. This cycle has been created for the same electric scooter as was used in Section 7.3.2 an approximation to the Peugeot SCOOT'ELEC, and is shown in Figure 7.12. The MATLAB® script file for creating this is very straightforward, and is given in Appendix 2. This cycle has the benefit that it probably models quite well the way such vehicles are used, considering the age of the typical rider, i.e. full speed for a good deal of the time!

There are many other test cycles which can be found in the literature, and some companies have their own in-house driving schedules. Academics sometimes propose new ones that, they suggest, better imitate real driving practice. There are also local driving cycles, which reflect the particular driving patterns of a city. A noteworthy example is the New York City Cycle, which has particularly long periods of no movement, and low average speeds, reflecting the state of the roads there. This cycle is sometimes used when simulating hybrid electric/ICE vehicles, as it shows this type of vehicle in a particularly good light.

The actual figures for the speed at each second, which are needed to run a simulation, can sometimes be deduced from the figures given above. However, in the case of the US cycles, which consist of a specific speed at each time, it is more convenient to load data files downloaded from web sites. These can readily be found using the normal internet search engines, though several have been supplied as MATLAB® script files at the website associated with this book.[4]

---

[4] www.wileyeurope.com/electricvehicles.

Electric Vehicle Modelling

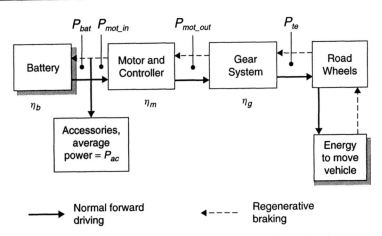

Figure 7.13 Energy flows in the 'classic' battery powered electric vehicle, which has regenerative braking

In the sections that follow it will be explained how a driving schedule can be simulated for different types of electric vehicle.

7.4.2 Range modelling of battery electric vehicles

7.4.2.1 Principles of battery electric vehicle modelling

The energy flows in a classical battery electric vehicle are shown in Figure 7.13. To predict the range the *energy* required to move the vehicle for each second of the driving cycle is calculated, and the effect of this energy drain is calculated. The process is repeated until the battery is flat. It is important to remember that if we use one-second time intervals, then the *power* and the *energy consumed* are equal.

The starting point in these calculations is to find the tractive effort, which is calculated from equation (7.9). The power is equal to the tractive effort multiplied by the velocity. Using the various efficiencies in the energy flow diagram, the energy required to move the vehicle for one second is calculated.

The energy required to move the vehicle for one second is the same as the power, so:

$$\text{Energy required each second} = P_{te} = F_{te} \times v \tag{7.23}$$

To find the energy taken from the battery to provide this energy at the road, we clearly need to be able to find the various efficiencies at all operating points. Equations that do this have been developed in the previous chapters, but we will review here the most important system modelling equations.

7.4.2.2 Modelling equations

The efficiency of the gear system η_g is normally assumed to be constant, as in electric vehicles there is usually only one gear. The efficiency is normally high, as the gear system will be very simple.

The efficiency of the motor and its controller are usually considered together, as it is more convenient to measure the efficiency of the whole system. We saw in Chapter 4 that motor efficiency varies considerably with power, torque, and also motor size. The efficiency is quite well modelled by the equation:

$$\eta_m = \frac{T\omega}{T\omega + k_c T^2 + k_i \omega + k_w \omega^3 + C} \quad (7.24)$$

where k_c is the copper losses coefficient, k_i is the iron losses coefficient, k_w is the windage loss coefficient and C represents the constant losses that apply at any speed. Table 7.2 shows typical values for these constants for two motors that are likely candidates for use in electric vehicles.

Table 7.2 Typical values for the parameters of equation (7.24)

Parameter	Lynch type PM motor, with brushes, 2–5 kW	100 kW, high speed induction motor
k_c	1.5	0.3
k_i	0.1	0.01
k_w	10^{-5}	5.0×10^{-6}
C	20	600

The inefficiencies of the motor, the controller and the gear system mean that the motor's power is not the same as the traction power, and the electrical power required by the motor is greater than the mechanical output power, according to the simple equations:

$$P_{mot\_in} = \frac{P_{mot\_out}}{\eta_m} \qquad P_{mot\_out} = \frac{P_{te}}{\eta_g} \quad (7.25)$$

Equations (7.25) are correct in the case where the vehicle is being driven. However, if the motor is being used to slow the vehicle, then the efficiency (or rather the inefficiency) works in the opposite sense. In other words the electrical power from the motor is reduced, and we must use these equations:

$$P_{mot\_in} = P_{mot\_out} \times \eta_m \qquad P_{mot\_out} = P_{te} \times \eta_g \quad (7.26)$$

So equations (7.25) or (7.26) are used to give use the electrical and mechanical power to (or from) the motor. However, we also need to consider the other electrical systems of the vehicle, the lights, indicators, accessories such as the radio, etc. An average power will need to be found or estimated for these, and added to the motor power, to give the total power required from the battery. Note that when braking, the motor power will be negative, and so this will reduce the magnitude of the power.

$$P_{bat} = P_{mot\_in} + P_{ac} \quad (7.27)$$

The meaning of these various powers, in and out of the motor, traction power and so on, is shown in Figure 7.13.

Electric Vehicle Modelling 203

The simulation of battery behaviour was explained in Section 2.11. To summarise, the procedure now is:

1. calculate the open circuit battery voltage, which depends on the state of charge of the battery;
2. calculate the battery current[5] using equation (2.20), unless P_{bat} is negative, in which case equation (2.22) should be used;
3. update the record of charge removed from the battery, correcting high currents using the Peukert coefficient, with equation (2.17); however, if the battery power is negative, and it is being charged, equation (2.23) should be used instead;
4. The level of discharge of the battery should then be updated, using equation (2.19).

Provided that the battery is not now too discharged, the whole process should then be repeated one second later, at the next velocity in the cycle.

7.4.2.3 Using MATLAB® or EXCEL® to simulate an electric vehicle

In the preceding section we saw how the various equations we have derived can be used to calculate what goes on inside an electric vehicle. To see how far a vehicle can go before the battery is flat we do this in a step-by-step process through the driving cycle. The way this is done is represented by the flowchart shown in Figure 7.14.

The first stage is to load the velocity data for the driving cycle to be used. This is usually done as by a separate MATLAB® script file. The way of doing this is explained in Appendix 2.

The next is to set up the vehicle parameters such as the mass, the battery size and type, and so on. The electrical power taken by the accessories P_{ac} should be set at this point.

Having done that, data arrays should be created for storing the data that needs to be remembered at the end of each cycle. These could be called 'end of cycle arrays'. The most important data that needs to be kept is a record of the charge removed from the battery, the depth of discharge of the battery, and the distance travelled.

The next stage is to set up arrays for the data to be stored just for one cycle; this data can be lost at the end of each cycle. This is also the charge removed, depth of discharge, and distance travelled, but we might also save other data, such as information about torque, or motor power, or battery current, as it is sometimes useful to be able to plot this data for just one cycle.

Having set the system up, the vehicle is put through one driving cycle, using the velocities given to calculate the acceleration, and thus the tractive effort, and thus the motor power, torque and speed. This is used to find the motor efficiency, which is used to find the electrical power going into the motor. Combined with the accessory power, this is used to find the battery current. This is then used to recalculate the battery state of charge. This calculation is repeated in one second steps[6] until the end of the cycle.

---

[5] If the vehicle uses a conventional DC motor, it might be more convenient to calculate the current using the more-or-less linear relationship between torque and current. If the connection is known, this can occasionally be a useful simplification.

[6] One second steps are the most convenient, as most driving cycles are defined in terms of one second intervals. Also, many of the formulas become much simpler. However, it is quite easy to adapt any of the programs given here for different time steps, and shorter steps are sometimes used.

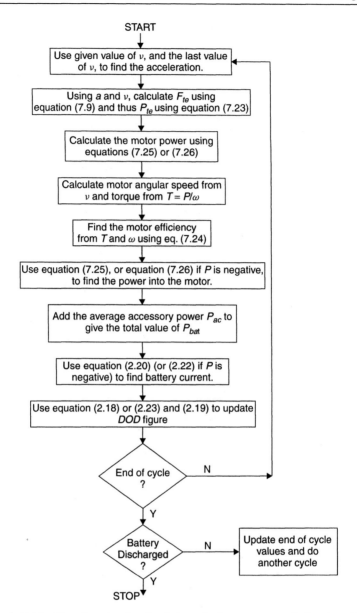

Figure 7.14 Flowchart for the simulation of a battery powered electric vehicle

The end of cycle data arrays are then updated, and if the battery still has enough charge, the process is repeated for another cycle. This is process is shown in the flowchart of Figure 7.14.

MATLAB® lends itself very well to this type of calculation. In Appendices 3 and 4 you will find example MATLAB® script files that find the range for a model of the

Electric Vehicle Modelling

famous General Motors EV1 vehicle. It should be easy enough to relate these to the text and all the equations given above. The main complications relate to zero values for variables such as speed and torque, which need careful treatment to avoid dividing by zero. The vehicle is running an urban driving cycle.

The file prints a graph of the depth of battery discharge against distance travelled, and this is shown in Figure 7.15, for two different situations. It can be seen that the vehicle range is about 130 km.

One of the very powerful features of such simulations is that they can be used to very quickly and easily see the effect of changing certain vehicle parameters on the range. For example, it is the work of a moment to change the program so that the conditions are different. For example, we can 'put the headlights on' by increasing the value of the average accessory power P_{ac}. We can also simulate colder weather by increasing the internal resistance by 25% or so, raising the Peukert coefficient and reducing the battery voltage very slightly. The simulation can then be re-run. This has been done with Figure 7.15. This shows how the depth of discharge rises under normal clement weather, daytime conditions, and also under colder conditions when in the dark. We can instantly see that the range, usually given as when 80% discharge is reached, drops from a little over 90 miles to about 70. The official stated range, in the GM literature, is '50 to 90 miles, depending on conditions.' Our simulation confirms this. We could further adapt the program to include hills, or more demanding driving, which would bring it below the 70 mile figure.

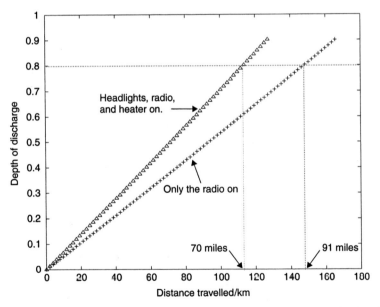

Figure 7.15 A graph of depth-of-discharge against distance travelled for a simulated GM EV1 electric car on the SFUDS driving cycle. In one case the conditions are benign, no lights, heating or air conditioning are in use. In the other case the battery is degraded slightly by cold weather, and all the vehicle headlights are on

The ECE-47 driving cycle was explained in the previous section. This can equally well be used for such range testing. In Appendix 5 we have given another MATLAB® script file for the same electric scooter that has been used for Figures 7.4 and 7.12. This vehicle has been set up with a NiCad battery, unlike the GMEV1, which uses lead acid. If the MATLAB® script file in Appendix 5 is studied, it will be seen that the vehicle has been fitted with three 100 Ah batteries, with the same properties as the NiCad batteries simulated in Section 2.11.3. Some range data, taken to 80% discharged, is given in Table 7.3.

Table 7.3 The simulated range of an electric scooter running the ECE-47 driving cycle with different degrees of regenerative braking

Percentage regenerative braking	Range on ECE-47 cycle
75% (not possible in practice)	51.30 km
50%	50.47 km
25%	49.59 km
None	48.82 km

The range of the scooter appears to be about 50 km, which is longer than the 40 km in 'urban nominal mode' claimed by the Peugeot SCOOT'ELEC, which uses the same batteries. This is probably due to the fact that, as we shall see in the next section, this ECE-47 driving cycle seems very well suited to the Lynch type motor we are using in our model. It may also be due to conservative claims in the vehicle specification.

Table 7.3 is also another demonstration of the power of simulations like these to quickly find the effect of changing vehicle parameters. In this case we have changed the proportion of the 'braking power' that is handled by the motor. In other words, we have changed the degree of regenerative braking performed. It is sometimes thought that this makes a huge difference to battery vehicle range. In the case of a scooter, it clearly does not. With no regenerative braking at all the range is 48.82 km. 50% is probably the highest practical possible motor braking, and this extends the range less than 2 km, or 4%. This does make some difference, but we should note that it is not a very great improvement. It is left as an exercise for the reader to do the same for the GM EV1; here the difference will be much greater, because it is a heavy vehicle, and well streamlined.

7.4.3 Constant velocity range modelling

Compared to the modelling of the driving cycles we have just achieved, constant velocity simulation is much easier. However, the basic round of calculations is the same as those outlined in the previous section. The system is simpler since the values of speed and torque are never negative or zero.

It would be possible to write a new and much shorter MATLAB script file for such a simulation. However, a quicker and easier solution, which makes use of the programs already written, is to create a 'driving cycle' in which the velocity is constant. This can be done in one line in MATLAB, thus:

```
linspace(12.5, 12.5, 100);
```

Electric Vehicle Modelling

This creates an array of 100 values, all equal to 12.5, which corresponds to 45 kph. A line like this, at any desired velocity, can replace the lines ECE_47, or SFUDS, at the beginning of the simulations given in the Appendices. This may not be the most elegant method, but it is probably the quickest. Constant velocity simulations are clearly very unrealistic, and so are of limited use.

7.4.4 Other uses of simulations

The data produced during these simulations has many more uses than just predicting the range of a vehicle. At each one second step of the cycle many variables were calculated, including:

- vehicle acceleration;
- tractive effort;
- motor power;
- motor torque;
- motor angular speed;
- motor power;
- motor efficiency;
- current out of (or into) the battery.

All of these variables are of interest, and it is instructive to plot them over one cycle. This can be done with great simplicity in MATLAB®, and it gives very useful results. The basic principle is to create two arrays with names such as XDATA and YDATA and allocate them values during a cycle. For example, if the loop counter is C then C will have the same value as the time in seconds. If we wanted to plot the value of the motor power during one cycle, then we would include the lines:

```
XDATA(C) = C;
YDATA(C) = Pmot_out;
```

into the code for a cycle. This can be seen very near the end of the script file given in Appendix 3. Near the end of the program, the main program that uses one_cycle, we would include the line:

```
plot(XDATA, YDATA);
```

For this type of plot the program could be simplified so that only ONE driving cycle is performed. An example of this type of plot output is shown in Figure 7.16. This shows the motor output power. It can be seen that the motor power is very modest, with a maximum of only about 12 kW. The motor has a maximum power of about 100 kW, so the simplified FUDS driving cycle is not testing the vehicle at all hard.[7]

---

[7] This can be confirmed by looking for the maximum acceleration during the SFUDS cycle, which is only about 1 ms^{-2}, whereas the GM EV1 is capable of over three times this value.

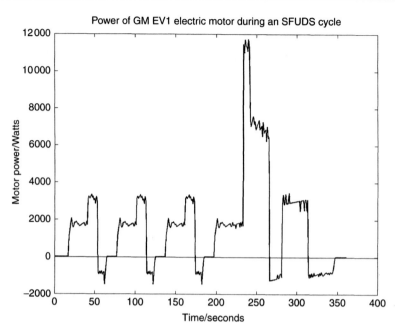

Figure 7.16 A graph of the power of the electric motor in a simulated GM EV1 electric car during one run of the SFUDS driving cycle

Another example of a particularly useful plot is that of motor torque against motor angular speed. To produce this graph only two lines of the program need changing. The X and Y data lines become:

```
XDATA(C) = omega;
YDATA(C) = Torque;
```

For this type of plot the points should be left as disconnected points – they should not be joined by a line. MATLAB easily allows this, and a suitable plot command is given near the end of the script file of Appendix 5. In Figure 7.17 this has been done for the electric scooter simulation. This maps the operating points of the motor. This graph should be carefully compared with Figure 6.7 in the previous chapter. It can be seen that, at least with this driving cycle, the motor is frequently operating in the region of about $120 \, \text{rad s}^{-1}$ speed, and low ($\sim 10 \, \text{Nm}$) torque. From Figure 6.7 we can see that this is *precisely the area where the motor is most efficient*. The motor is thus extremely well matched to this particular driving cycle. This probably explains why the range simulation results were rather better that given in the specification for Peugeot SCOOT'ELEC, to which our model is quite similar.

7.4.5 Range modelling of fuel cell vehicles

The principal energy flows in a fuel cell powered vehicle are shown in Figure 7.18. The energy required to drive the various fuel cell ancillaries that were discussed in Chapter 4

Electric Vehicle Modelling

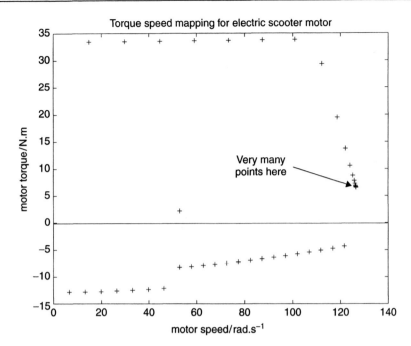

Figure 7.17 A plot of the torque/speed operating points for the electric motor in an electric scooter during the ECE-47 test cycle. In the indicated region, many points are superimposed, as the vehicle is at a constant velocity

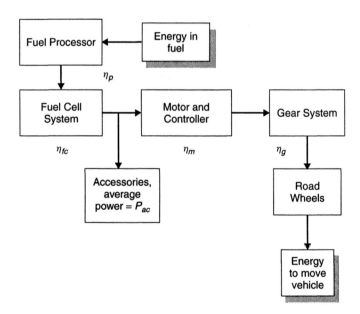

Figure 7.18 Energy flows in a fuel cell powered electric vehicle

have not been explicitly shown, but these can be accounted for by adjustments to the value of the fuel cell efficiency.

The modelling of such a system is extremely complex, largely because of the fuel processor system. This has very many sub-processes with highly variable time constants, some quite long. The simulation of such fuel processing systems is extremely important, but too complex for an introductory text such as this. In addition, most of the important data is highly confidential to the companies developing these systems.

However, the simulation of a system running directly off onboard stored hydrogen is not nearly so complex. Indeed in many ways it can be less difficult that for battery vehicles, at least to a first approximation. The efficiency of a fuel cell is related, as we saw in Chapter 4, to the average voltage of each cell in the fuel cell stack V_c. If the efficiency is referred to the lower heating value (LHV) of hydrogen, then:

$$\eta_{fc} = \frac{V_c}{1.25} \qquad (7.28)$$

Now, we know from Chapter 4 that at lower currents the fuel cell voltage rises, and thus also the efficiency. However, we also saw in Chapter 4 that a fuel cell system will also have many pumps, compressors, controllers and other 'balance of plant' that use electrical power. This use of electrical power is higher, as a proportion of output power, at lower currents. The result is that, in practice, the efficiency of a fuel cell is, to quite a good approximation, more-or-less constant at all powers. (Note, this contrasts with an IC engine, whose efficiency falls very markedly at lower powers.)

At the present time, a good target value for the efficiency of a fuel cell operating off pure hydrogen is 38% referred to the LHV. So, from equation (7.28), we have:

$$V_c = 0.38 \times 1.25 = 0.475 \text{ V} \qquad (7.29)$$

Note that the fuel cell will probably, in fact, be running at about 0.65 V, but the difference between this and 0.475 represents the energy used by the balance of plant.

This value of average cell voltage can then be used in the formula[8] for the rate of use of hydrogen in a fuel cell:

$$\text{H}_2 \text{ rate of usage } \dot{m} = 1.05 \times 10^{-8} \times \frac{P}{V_c} = 2.21 \times 10^{-8} \times P \qquad (7.30)$$

Notice that this formula does not require us to know any details about the fuel cell, such as the number of cells, electrode area, or any details at all. It allows us to very straightforwardly calculate the mass of hydrogen used each second from the required electrical power. Indeed, this simulation is a great deal easier than with batteries because:

- there is no regenerative braking to incorporate;
- there are no currents to calculate;
- there is no Peukert correction of the current to be done.

---

[8] Derived in Appendix 2 of Larminie and Dicks (2003).

Electric Vehicle Modelling

By way of example, we could take our GM EV1 electric vehicle, and take out the 594 kg of batteries. In their place we could put the fuel cell system shown in Figure 4.23, and the hydrogen storage system outlined in Table 5.3. The key points are:

- mass of hydrogen stored = 8.5 kg;
- mass of storage system = 51.5 kg;
- mass of 45 kW fuel cell system = 250 kg (estimate, not particularly optimistic);
- total mass of vehicle is now $(1350 - 594) + 8.5 + 51.5 + 250 = 1066$ kg.

Appendix 6 gives the MATLAB script file for running the SFUDS driving cycle for this hypothetical vehicle. It can be seen that the simulation is simpler. Some example results, *which the reader is strongly encourage to confirm*, are given in Table 7.4 below. In both cases 80% discharge is taken as the end point, i.e. 1.7 kg of hydrogen remaining in the case of the fuel cell.

An alternative approach that might well be found helpful, that certainly results in a much simpler MATLAB program, is to compute the energy consumed in running one cycle of the driving schedule being used. The distance travelled in one cycle should also be found. The number of cycles that can be performed can then be computed from the available energy and the overall efficiency. This approach obviously gives the same result.

7.4.6 Range modelling of hybrid electric vehicles

All the modelling we have done so far has involved equations which the system followed in a more-or-less predetermined way. However, when we come to a hybrid electric vehicle, then this is no longer so. Hybrid electric vehicles involve a controller, which monitors the various power in the system, and the state of charge of the battery, and makes decisions about the power to be drawn from the engine, battery, and so on. Very little about the energy flow is inevitable and driven by fixed equations. Furthermore, the strategy will change with time, depending on issues such as when the vehicle was last used, the temperature, the need to equalise the charge in the batteries from time to time, and a host of other criteria. The decision making of these controllers is not at all simple to simulate.

Another complication is that there are so many different configurations of hybrid electric vehicle.

The result is that the simulation of these vehicles cannot be attempted or explained in a few pages of a book like this. Indeed, the use of a simple program such as MATLAB® on its own is probably not advisable. At the very least the SIMULINK® extension to MATLAB® should be used. There are a number of vehicle simulation programs available

Table 7.4 Simple fuel cell simulation results, showing great improvement in range over a battery vehicle

Simulation, SFUDS driving cycle in both cases	Range
GM EV1 with standard lead acid batteries, good conditions	148 km
GM EV1 with a fuel cell and cryogenic H_2 store.	485 km

or described in the literature, for example Bolognesi *et al.* (2001). Among the most well known of these is ADVISOR® (Wipke *et al.* 1999), which is MATLAB®-based.

The program for the energy controller in a hybrid system, which is a sub-system of vital importance, will often be written in a high level language such as C. It makes sense to incorporate the simulator in the same language, and then it should be possible to use the very same control program that is being written for the controller in the simulation. This has obvious reliability and efficiency benefits.

7.5 Simulations: a Summary

We have seen in this chapter how to begin the simulation or modelling of the range and performance of electric vehicles. With most vehicles the simulation of performance, by which we usually mean acceleration, is fairly straightforward. Mathematical software such as MATLAB® or EXCEL® lend themselves very well to this.

In the case of the classical battery powered electric vehicle, and fuel cell vehicles using stored hydrogen, the modelling of range, though considerably more complex, is not difficult. For hybrid vehicles a great deal of care and thought is needed in setting up a simulation.

We have shown the fact that a vehicle mathematical model is crucially important in the design of electric vehicles, as it allows the designer very quickly to try out different design options, and at virtually no cost. By using examples of the GM EV1 electric car, and an electric scooter, we have shown how even a quite simple mathematical model is validated by the performance of the real vehicle.

It is almost impossible to over-emphasise the importance of this modelling process in electric vehicle design. Hopefully these explanations, and the example MATLAB® script files we have given, will enable you to make a start in this process. Chan and Chau (2001) have described the use of an extension of this sort of programming in their simulation program EVSIM, and this could be studied to see how these ideas could be taken further.

References

Bolognesi P., Conte F.V., Lo Bianco G. and Pasquali M. (2001) Hy-Sim: a modular simulator for hybrid-electric vehicles. *Proceedings of the 18th International Electric Vehicle Symposium*, (CD-ROM).
Chan C.C. and Chau K.T. (2001) *Modern Electric Vehicle Technology*, Oxford University Press, Oxford.
Larminie J. and Dicks A. (2003) *Fuel Cell Systems Explained*, 2nd Edn, Wiley, Chichester.
Shnayerson M. (1996) *The Car that Could*, Random House, New York.
Wipke K., Cuddy M., Bharathan D., Burch S., Johnson V., Markel A. and Sprik S. (1999) ADVISOR 2.0: A Second Generation Advanced Vehicle Simulator for Systems Analysis, NREL Report no. TP-540-25928, Golden, CO (http://www.ctts.nrel.gov/analysis/).

8

Design Considerations

8.1 Introduction

To maximise the fuel efficiency of any vehicle the mass, aerodynamic drag and rolling resistance have to be minimised, while at the same time maximising engine/motor and transmission efficiencies. It is particularly important to design battery electric vehicles with high efficiencies in order to reduce the mass of expensive batteries required.

This chapter builds on the previous chapter on vehicle modelling. The various parameters that went into the model are examined individually, together with their effect on vehicle performance. Various choices available to designers to optimise their vehicle design are discussed, as is the greater flexibility to place components in electrical vehicles with a view to optimise weight positioning and minimise aerodynamic drag.

8.2 Aerodynamic Considerations

8.2.1 Aerodynamics and energy

It is well known that the more aerodynamic is a vehicle, the lower is its energy consumption. Bearing in mind the high cost of onboard electric energy, the aerodynamics of electric vehicles is particularly important, especially at high speeds.

Let us first consider the effect of aerodynamic drag. As seen in Chapter 7, the drag force F_{ad} on a vehicle is:

$$F_{ad} = \tfrac{1}{2}\rho A C_d v^2 \tag{8.1}$$

and the power P_{adw} (Watts) at the vehicle wheels required to overcome this air resistance is:

$$P_{adw} = F_{adw} \times v = \tfrac{1}{2}\rho A C_d v^3 \tag{8.2}$$

where ρ is density of air (kg.m^{-3}), A is the frontal area (m^2), v is the velocity (m.s^{-1}) and C_d is the drag coefficient, which is dimensionless.

Electric Vehicle Technology Explained   James Larminie and John Lowry
© 2003 John Wiley & Sons, Ltd   ISBN: 0-470-85163-5

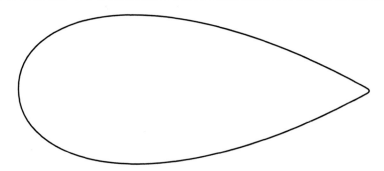

Figure 8.1 Aerodynamic idea shape, a teardrop of aspect ratio 2.4

The ideal aerodynamic shape is a teardrop, as achieved by a droplet of water freefalling in the atmosphere as illustrated in Figure 8.1. The coefficient of drag varies with the ratio of length to diameter, and it has a lowest value of $C_d = 0.04$ when the ratio of the length to diameter is 2.4. Using equation (8.2) and taking air density to be $1.23 \, \text{kg.m}^{-3}$, the power required to drive a teardrop shaped body with $C_d = 0.04$ of cross-section $1 \, \text{m}^2$ travelling at 100 kph ($27.8 \, \text{m.s}^{-1}$) in clear air will be 664 W. If engineers and scientists could achieve such aerodynamic vehicle shapes they would revolutionise energy in transport. Unfortunately they cannot get near such a low value. However, the ideal teardrop shape is normally an 'aiming point' for vehicle aerodynamicists.

In reality the drag coefficients of vehicles are considerably higher due to various factors, including the presence of the ground, the effect of wheels, body shapes which vary from the ideal, and irregularities such as air inlets and protrusions.

The aerodynamic drag coefficient for a saloon or hatchback car normally varies from 0.3 to 0.5, while that of reasonably aerodynamic van is around 0.5. For example, a Honda Civic hatchback has a frontal area of $1.9 \, \text{m}^2$ and a drag coefficient of 0.36. This can be reduced further by careful attention to aerodynamic detail. Good examples are the Honda Insight hybrid electric car, with a C_d of 0.25, and the General Motors EV1 electric vehicle with an even lower C_d of 0.19. The Bluebird record-breaking electric car had a C_d of 0.16 (A sphere has a C_d of 0.19.)

As the drag, and hence the power consumed, is directly proportional to the drag coefficient, a reduction from a C_d of 0.3 to a C_d of 0.19 will result in a reduction in drag of 0.19/0.3, i.e. 63.3%. In other words the more streamlined vehicle will use 63.3% of the energy to overcome aerodynamic drag compared to the less aerodynamic car. For a given range, the battery capacity needed to overcome aerodynamic resistance will be 36.7% less. Alternatively the range of the vehicle will be considerable enhanced.

The battery power P_{adb} needed to overcome aerodynamic drag is obtained by dividing the overall power delivered at the wheels P_{adw} by the overall efficiency η_0 (power at wheels/battery power).

$$P_{adb} = \frac{P_{adw}}{\eta_0} = \frac{\frac{1}{2}\rho A C_d v^3}{\eta_0} \tag{8.3}$$

The battery mass m_b (kg) of a battery with specific energy SE (Wh.kg^{-1}) required to overcome the aerodynamic drag at a velocity v (ms^{-1}) over a distance d (metres) is

Design Considerations

given by:

$$m_b = \frac{P_{adb} \times d}{v \times SE \times 3600} \text{ kg} \quad (8.4)$$

The variation of battery power P_{adb} for overcoming aerodynamic drag with speed is shown in Figure 8.2 for vehicles of different drag coefficients and different frontal areas. The battery mass required to provide energy to overcome aerodynamic drag for a vehicle with a range of 100 km travelling at different constant speeds is shown in Figure 8.3. An efficiency η_0 of 0.7 is used. Figure 8.3 dramatically illustrates the importance of streamlining, as the battery weight shown in this graph is purely that needed to overcome wind resistance, and for the not very impressive range of 100 km. Figure 8.3 also clearly shows how ill-suited battery electric vehicles are to high-speed driving. Even a well designed car, with a C_d of 0.19, still needs about 400 kg of lead acid batteries to travel just to overcome wind resistance for 100 km when going at 160 kph. If the MATLAB® file used in Section 7.4 for the range modelling of the GM EV1 (whose results are shown in Figure 7.15) are adapted for a constant speed of 120 kph, it will be found that the predicted range is less than 80 km However, when driving the SFUDS cycle, which has

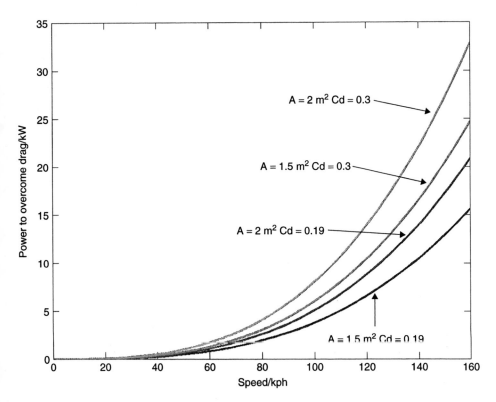

Figure 8.2 Power requirement to overcome aerodynamic drag for vehicle of different frontal areas and drag coefficients for a range of speeds up to 160 kph

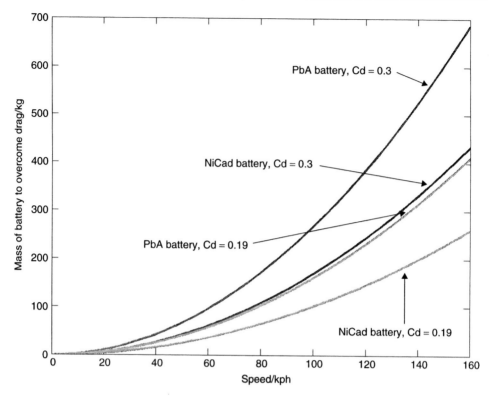

Figure 8.3 The effect of drag coefficient and speed on battery mass. The vehicles all have a frontal area of $1.5\,\text{m}^2$, and the range is 100 km. The mass is only for the energy required to store the energy to overcome aerodynamic drag; the actual battery mass would need to be higher

plenty of stopping and starting, but no speeds over 60 kph, the range could be over 140 km in good conditions.

It is clear from both Figures 8.2 and 8.3 that there are huge advantages in keeping both the aerodynamic resistance and the vehicle frontal area as low as possible. Bearing in mind the considerable cost saving on both battery weight and battery cost it is well worth paying great attention to the aerodynamic details of the chassis/body. There is great scope for producing streamlined shapes with battery electric vehicles as there is much more flexibility in placing major components and there is less need for cooling air ducts and under vehicle exhaust pipes. Similarly as well as keeping the coefficient of drag low, it is equally important to keep down the frontal area of the vehicle if power requirements are to be minimised. Although the car needs to be of sufficient size to house the passengers in comfort, the greater flexibility in which components can be placed in an electric car can be used to minimise frontal area.

Some consideration also needs to be given to items such as wing mirrors, aerials and windscreen wipers. These need to be designed to minimise the drag. Aerials do not need to be external to the car body and wing mirrors can be replaced by electronic video systems that can be contained within the aerodynamic envelope of the car. Whilst the latter may appear an expensive option at first, the reduced drag will result in a lighter battery with associated cost savings.

8.2.2 Body/chassis aerodynamic shape

The aerodynamic shape of the vehicle will depend largely on the type of use to which the electric vehicle is to be put. If it is a city commuter car or van that will be driven at relatively low speeds, the aerodynamics are much less important than on a conventional vehicle which will be used for motorway driving.

For the latter type of vehicle, to be used at relatively high speeds, a low frontal area and streamlining is vitally important. It is worth having a look at how this was achieved with vehicles such as the Honda Insight hybrid ($C_d = 0.25$) and the GM EV1 battery car ($C_d = 0.19$). Although the aerodynamics of high speed battery electric vehicles are vitally important, they are also important for hybrid vehicles, but optimisation can result in a slightly less aerodynamic vehicle with more reliance being placed on the internal combustion engine to achieve range.

Most aerodynamic vehicles at least attempt to copy the teardrop shape and this is true of both these two vehicles. The body shape is also designed to keep the airflow around the vehicle laminar.

On the Honda Insight, shown in Figure 8.15, the body is tapered so that it narrows towards the back, giving it a shape approaching the teardrop. The rear wheels are placed approximately 110 mm closer together than the front wheels, allowing the body to narrow. The cargo area above the wheel wells is narrower still, and the floor under the rear portion of the car slopes upwards, while the downward slope of the rear hatch window also contributes to the overall narrowing of the car at the rear.

At the back of the Insight the teardrop shape is abruptly cut off in what is called the Kamm effect. The Kamm back takes advantage of the fact that beyond a certain point there is little aerodynamic advantage of rounding off or tapering, so it might as well be truncated at this point avoiding long, extended, fairly useless tail sections.

Another important aerodynamic feature is the careful management of under-body airflow. The Insight body features a flat under-body design that smoothes airflow under the car, including three under-body covers. Areas of the under-body that must remain open to the air such as the exhaust system and the area under the fuel tank (it is a hybrid) have separate fairings to smooth the flow around them.

In order to minimise the air leakage to the underside the lower edges of the sides and the rear of the body form a strake that functions as an air dam. At the rear the floor pan rises at a five degree angle toward the rear bumper, creating a gradual increase in the body area that smoothly feeds under-body air into the low pressure area at the rear of the vehicle.

The GM EV1 with its exceptionally low drag adopts a similar approach. It has the advantage that as a pure electric vehicle there is no need for fuel tanks or exhaust pipes. Again the vehicle shape emulates as far as possible the teardrop shape and is as perfectly smooth as possible. The rear wheels on the EV1 are 228 mm closer together, nearly twice that on the Insight. This can clearly be seen in Figure 11.5, which shows several views of this vehicle. With both vehicles abrupt changes in body curvature are avoided, items such as the windscreen are joined smoothly into the shape, and gaps such as that between the wheels and body are minimised. The surface of the wheels blends in with the body shape. This all helps to keep the airflow laminar and thus reduce drag.

On very low speed vehicles (<30 kph), such as golf buggies, where tranquility and pure air are more important than rushing around at speed, the aerodynamic shape of the vehicle is almost irrelevant. As discussed earlier, with vehicles such as commuter cars and town delivery vehicles the aerodynamic shape is less important than on faster cars. However, it does have some significance and should not entirely be ignored. There have been attempts to produce both vans and buses with tear drop shapes, but there is a conflict between an aerodynamic shape and low frontal area and the need for maximising interior space, particularly for load-carrying. There is nothing stopping commuter vehicles from being aerodynamic but in the case of vans there needs to be a compromise between the need to slide large items into a maximum space and the desire for a teardrop tail. There is no reason why some of the features used on the Insight and the EV1, such as under-body covers, cannot be used on vans. Careful consideration of the van shape, where possible avoiding rapid changes in curvature, keeping the wheel surfaces flat, minimising gaps and rounding the corners will indeed reduce the drag coefficient, but not down to that of the EV1.

When carrying out initial calculations, as in a feasibility study, the coefficient of drag is best estimated by comparing the proposed vehicle with one of a similar shape and design. Modern computational fluid dynamics (CFD) packages will accurately predict aerodynamic characteristics of vehicles. Most motor manufacturers use wind tunnels either on scale models or more recently on full size vehicles. Some of these now incorporate rolling roads so that an almost exact understanding of drag, lift, etc., can be measured.

8.3 Consideration of Rolling Resistance

As discussed in Chapter 7, the rolling drag on a vehicle F_{rr} is given by:

$$F_{rr} = \mu_{rr} mg \tag{8.5}$$

where μ_{rr} is the coefficient of rolling resistance. The rolling drag is independent of speed. Power needed to overcome rolling is given by:

$$P_{rr} = F_{rr} \times v = \mu_{rr} mg v \tag{8.6}$$

The value of μ_{rr} varies from 0.015 for a radial ply tyre down to 0.005 for tyres specially developed for electric vehicles. A reduction of rolling resistance to one-third is a substantial benefit, particularly for low speed vehicles such as buggies for the disabled. For low speed vehicles of this type the air resistance is negligible and a reduction of drag to one-third will either triple the vehicle range or cut the battery mass and cost by one-third, a substantial saving both in terms of cost and weight.

Power requirements/speed for an electric vehicle travelling on the flat, with typical drag ($C_d = 0.3$) and fairly standard tyres ($\mu_{rr} = 0.015$), with a mass of 1000 kg and a frontal

Design Considerations

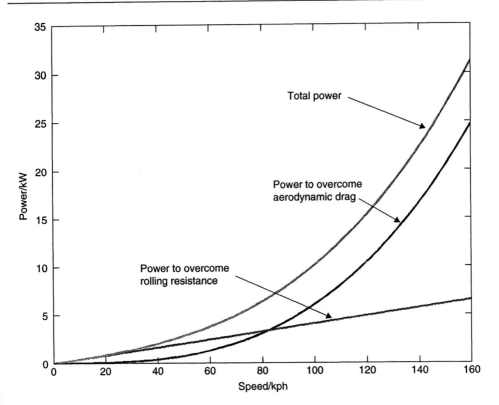

Figure 8.4 The power requirements to overcome rolling resistance and aerodynamic drag at different speeds. This is for a fairly ordinary small car, with $C_d = 0.3$, frontal area $1.5\,\text{m}^2$, mass = 1000 kg, and $\mu_{rr} = 0.015$

area of $1.5\,\text{m}^2$ is shown in Figure 8.4. The graph, derived from the above equations, shows how much power is required to overcome rolling resistance and aerodynamic drag.

It can be seen clearly in Figure 8.4 that at low speeds, e.g. under 50 kph, aerodynamics have very little influence, whereas at high speeds they are the major influence on power requirements. It may be concluded that streamlining is not very important at relatively low speeds, more important at medium speeds and very important at high speeds. So, for example, on a golf cart the aerodynamics are unimportant, whereas for a saloon car intended for motorway driving the aerodynamics are extremely important. (The rolling resistance of a golf buggy wheels on turf will of course be considerably higher than can be expected on hard road surfaces.)

A graph of the total power requirement for two vans is shown in Figure 8.5, where a power/velocity curve for each vehicle is plotted. Both vans have a mass of 1000 kg, frontal area of $2\,\text{m}^2$ and a C_d of 0.5. However one has ordinary tyres with a μ_{rr} of 0.015, whereas the other has low rolling resistance tyres for which μ_{rr} of 0.005.

It can be concluded that for all electric vehicles a low rolling resistance is desirable and that the choice of tyres is therefore extremely important. A low coefficient of aerodynamic

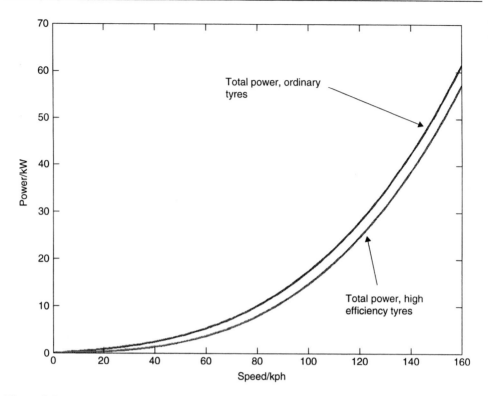

Figure 8.5 Power requirements for aerodynamic drag and rolling resistance at a range of speeds. This is for two vans, both of mass 1000 kg, frontal area 2 m², and $C_d = 0.5$. However, one has low resistance tyres with $\mu_{rr} = 0.005$, whereas the other has ordinary tyres for which $\mu_{rr} = 0.015$

drag is very important for high speed vehicles, but is less important for town/city delivery vehicles and commuter vehicles. On very low speed vehicles such as electric bicycles, golf buggies and buggies for the disabled, aerodynamic drag has very little influence, whereas rolling resistance certainly does.

8.4 Transmission Efficiency

All vehicles need a transmission that connects the output of the motor to the wheels. In the case of an internal combustion engine vehicle the engine is connected to a clutch which in turn connects to a gear box, a prop shaft, a differential (for equalising the torque on the driving wheels) and an axle.

All of these have inefficiencies that cause a loss of power and energy. The transmission of electric vehicles is inherently simpler than that of IC engine vehicles. To start with no clutch is needed as the motor can provide torque from zero speed upwards. Similarly, a conventional gear box is not needed, as a single ratio gear is normally all that is needed. The three basic variations of electric vehicle transmission are illustrated in Figure 8.6.

Design Considerations

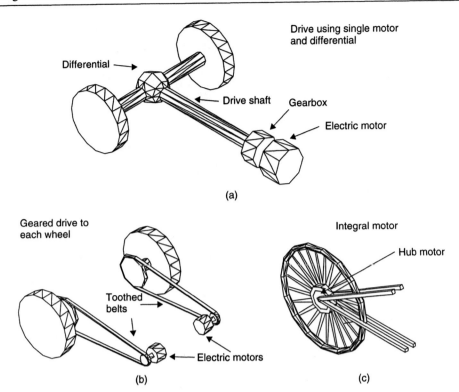

Figure 8.6 Three different arrangements for electric vehicle transmission

The most conventional arrangement is to drive a pair of wheels through a differential. This has many advantages, the differential being a well tested, reliable, quantity-produced piece of engineering. The disadvantage is that some power is lost through the differential, and differentials are relatively heavy. It can also take up space in areas where the space can be usefully utilised. An example of a motor and differential fitted to an experimental battery powered vehicle is shown in Figure 8.7. In this system the motor is transversal, but otherwise it is similar to Figure 8.6. This arrangement can also be seen in the diagram of the electric bus in Figure 11.7.

The differential can be eliminated by connecting a motor to each wheel via a single ratio gearbox or even a toothed belt drive. The torque needs to be equalised to each wheel by the electronic controller. This system has the advantage of clearing space within the vehicle, and the disadvantage of needing a more complicated electronic controller. Also, in terms of cost per kilowatt, two small motors are considerably more expensive than one larger one. An example of a small motor connected via a simple gearbox to an axle, which would be suitable for this sort of application, is shown in Figure 8.8.

The third method is to directly connect the motor to the wheels via a shaft, or to actually design the motor as part of the hub assembly. This system has huge potential advantages, including a 100% transmission efficiency. The trouble with this system is that most electric motors typically run at 2 to 4 times faster than the vehicle wheels, and designing a

Figure 8.7 Example of type (a) of Figure 8.6 on an experimental electric vehicle by MES-DEA of Switzerland. The mounting of the motor is transverse, so there is no drive shaft

Figure 8.8 Commercial motor and single speed gearbox to axle connection. This type of motor is designed for use in systems like that of Figure 8.6(b), or on vehicles with a single drive wheel, or on vehicles like go-karts which have no differential

motor to work slowly results in a large heavy motor. However, this arrangement has and can be used. It is particularly popular in electric scooters and bicycles. An example is shown in Figure 8.9. The General Motors Hy-wire of Figure 8.16 uses this approach, and it can also be seen in the electric bicycle of Figure 11.1. Normally a vehicle's handling is improved if the unsprung mass is kept to a minimum. Placing the motor in the hub has

Design Considerations

Figure 8.9 The rear wheel of an EVT electric scooter. Here there is no transmission, the wheel and motor are one. This is an example of Figure 8.6(c)

advantages for space saving in vehicle layout, but will adversely affect handling. Also, the motor is certain to be considerably more expensive in terms of cost per kilowatt.

Of course, if you were designing a three-wheeler and driving the single wheel you would not need any differential, mechanical or electronic! You may still need to gear the motor to the wheel. A tricycle arrangement with one driven wheel at the back could also help in the production of a near teardrop shape with its associated low aerodynamic drag. Such an arrangement has been used in some experimental vehicles. A power unit that could be the basis for such a vehicle is shown in Figure 8.11.

Whatever the arrangement for the transmission, the transmission efficiency is important. A 10% increase in transmission efficiency will allow a similar reduction in battery mass and battery cost or alternatively a 10% increase in the vehicle range.

8.5 Consideration of Vehicle Mass

The mass of an electrical vehicle has a critical effect on the performance, range and cost of an electric vehicle. The first effect of the mass on rolling resistance and the power and energy to overcome this has already been discussed in Section 8.3.

There are two other effects of mass. The first concerns a vehicle climbing a hill and the second is the kinetic energy lost when the vehicle is accelerating and decelerating in an urban cycle.

In Chapter 7 equation (7.3) it was seen that the Force F_{hc} in Newtons along the slope for a car of mass m (kg) climbing a hill of angle ψ is given by:

$$F_{hc} = mg \sin \psi \tag{8.7}$$

It follows that the power P_{hc} in Watts for a vehicle climbing a slope at a velocity v ms^{-1} is given by:

$$P_{hc} = F_{hc} \times v = mgv \sin \psi \tag{8.8}$$

Figure 8.10 shows the *total* power needed to travel at a constant 80 kph up slopes of varying angles up to $10°$ for vehicles of two different weights, but otherwise similar. They are based loosely on the GM EV1 electric car studied in Chapter 7. They both have a drag coefficient of 0.19, and tyres with coefficient of rolling resistance of 0.005, and the frontal area is 1.8 m². We can see that the 1500 kg car, which is approximately the weight of the real GM EV1, has to provide approximately *12 times as much power* at 10° than is

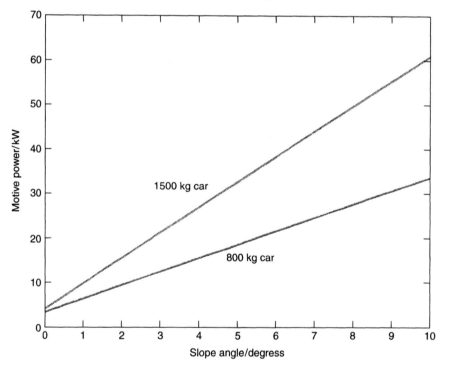

Figure 8.10 The total power requirements for two different vehicles moving at 80 kph up a hill of slope angle zero to 10°. In both cases the vehicle has good tyres with $\mu_{rr} = 0.005$, low drag as $C_d = 0.19$, and a frontal area of 1.8 m². One car weighs 800 kg, the other 1500 kg

needed on the flat. With the 800 kg vehicle the power needed increases greatly, but only by about 8 times.

Looking at Figure 8.10 we see why the GM EV1 electric car needs a motor of power about 100 kW. In the SFUDS simulation we noted that the maximum power needed was only 12 kW, as in Figure 7.16. It is taking heavy vehicles up hills that requires high power.

The results shown in the graph send a clear message. Considerable power is required for hill climbing, and such terrain will restrict the range of electric vehicles relying solely on rechargeable batteries. When designing electric vehicles the effect of hills must be taken into account, though there are no agreed 'standard hills' for doing this. It is not too difficult, after a little experience, to add gradients to the simulation driving cycles considered in the last chapter. This is usually done with a specific journey in mind.

The effect of the vehicle mass when accelerating and stopping in town and city conditions is another area where the mass of the electric vehicle will have considerable influence on vehicle performance. There are a variety of simulated urban driving cycles that have already been discussed in Chapter 7. Basically when a vehicle of mass m (kg) is travelling at velocity v (ms^{-1}) its kinetic energy is given by:

$$KE = \tfrac{1}{2} mv^2 \tag{8.9}$$

If the vehicle brakes this energy is converted into heat. When regenerative braking is used a certain amount of the energy is recovered. This was extensively explored in Chapter 7, Section 7.4.2.3 and Table 7.3. The maximum practical limit on the recovery of kinetic energy is about 40%. In light vehicles the losses associated with continually creating and then losing kinetic energy are much less, and the benefits of regenerative braking are similarly reduced.

Apart from the importance of minimising vehicle weight, it is also important to try to minimise the moment of inertia of rotating components, as these store rotational kinetic energy. The energy stored E_r (Joules) of a component with a moment of inertia I (kg.m^2) rotating at ω (rad s^{-1}) is given by:

$$E_r = \tfrac{1}{2} I \omega^2 \tag{8.10}$$

The moment of inertia I is normally expressed as:

$$I = \sum_{n=1}^{n=N} m_n r_n^2 \tag{8.11}$$

i.e. the sum of all the finite masses of a component which lie a distance r from the centre of rotation. In practice most rotating components such as the wheels are purchased as proprietary items, but the energy lost in rotary energy needs to be considered particularly for urban driving conditions. This was addressed in Section 7.2 and equation (7.8). In practice it is often difficult to obtain precise information about the moment of inertia of the rotating parts, and a reasonable approximation is to simply increase the mass in equation (8.9) by 5%, and not use equation (8.10). Notice that this does not need to be done for the mass in the hill climbing equation (8.8).

In the next section we consider the chassis and body design, and how it might be made, and what materials used, in order to achieve this aim of reducing the weight.

8.6 Electric Vehicle Chassis and Body Design

8.6.1 Body/chassis requirements

This section is intended to give guidance on the design of chassis for electric vehicles. Chassis design should be carried out in conjunction with other texts on chassis design, not to mention computer packages that specialise in this area. Nevertheless a basic understanding of what the chassis should do and other parameters related to electric vehicle chassis is needed.

In the early cars chassis and bodies were separate items. The chassis gave the basic strength of the vehicle while the body and glazing acted as a cocoon to keep the passengers and luggage protected from the outside elements.

In recent times the body and chassis have been combined as a monocoque so that every part, including the glazing, adds to the strength and stiffness, resulting in a much lighter vehicle. Either monocoques or separate chassis/body units are an acceptable basis for design. Despite the popularity of monocoques, several modern electric vehicles use a separate chassis, most notably the advanced new General Motors Hy-wire fuel cell vehicle, which will be discussed in more detail later.

It is worth pausing to think precisely what the chassis/body does; ideally a chassis/body should fill the following criteria. It should:

- be strong;
- be light;
- be rigid;
- not vibrate, particularly at frequencies and harmonics of rotating parts;
- be aerodynamic;
- be resistant to impact;
- crumple evenly in an accident, minimising forces on driver/passengers;
- be strong enough to fix components to easily;
- be impact and roll resistant;
- be cheap;
- be aesthetically pleasing;
- be corrosion-proof.

Chassis/body design requires optimisation of conflicting requirements such as cost and strength, or performance and energy efficiency. There are important differences when designing electric vehicles compared to their IC equivalents. For example, extra weight is not so important with an internal combustion vehicle, where a little more power can be cheaply added to compensate for a slightly heavier chassis. The same is true for aerodynamic drag, where a slight increase in drag can be similarly compensated for. Savings in weight as well as increases in efficiency contribute directly to the size of the batteries and these are both heavy and expensive.

Design Considerations

It must also be borne in mind that most internal combustion engine vehicles are quantity produced, whereas at the moment, and probably for the immediate future, small scale production of electric vehicles is likely. This in itself will tend to result in the use of materials such as reinforced plastics, where there is potential scope for more perfect aerodynamic shapes and weight saving.

8.6.2 Body/chassis layout

There is plenty of scope for designers of electric vehicles to experiment with different layouts to optimise their creation. To start with there is no need for a bonnet housing and engine. In addition, batteries can be placed virtually anywhere along the bottom (for stability) of the vehicle and motors and gearing can be, if required, integrated with the wheel hub assemblies.

Most batteries can be varied in size. Height can be traded against length and width, and most batteries (not all) can be split up so that they can be located under seats and anywhere else required, all of which can help to use every available space and to reduce the vehicle frontal area. Batteries can also be arranged to ensure that the vehicle is perfectly balanced around the centre of gravity, giving good handling characteristics.

A picture of an interesting experimental drive system assembly is shown in Figure 8.11, consisting of one driven wheel, with batteries and controller all built into the unit. The scope for using such a device on a range of interesting vehicle layouts is considerable. It could be incorporated, for example, to drive the rear wheel in a tricycle arrangement. Interestingly one of the most popular three-wheel electric vehicles is the Twike illustrated in Figure 8.12. Based on the previous argument the vehicle layout could be interpreted

Figure 8.11 This power module comprises motor, controller, battery and one driven wheel in a neat unit that could be built into a wide range of vehicle designs

Figure 8.12 The famous Swiss electric Twike

as being the 'wrong way round'; the body is like a tear drop going backwards. However, as it is a low-speed commuter vehicle based on bicycle components the aerodynamic shape is not as important as those of a high-speed vehicle. The two rear wheels with the passengers sitting side by side give stability.

The layout for an electric van also has considerable scope for new ideas. Electric motors and gearing can again, if required, be incorporated into the wheel hub assemblies, avoiding space requirements for motors, gearing and transmission. Batteries such as lead acid, NiCad or NiMH can be spread as a thin layer over the base of the vehicle leaving a large flat floored area above, an essential requirement for vans.

8.6.3 Body/chassis strength, rigidity and crash resistance

The days have long past, thankfully, when stress engineers regarded aircraft as hollow cylinders with beams stuck out of the side and cars as something simpler. Modern predictions of chassis body behaviour and virtually every aspect of car design rely ultimately on complex computer packages. Nevertheless a basic understanding of the behaviour of beams and hollow cylinders does give an insight into body chassis design.

Let us look at a hollow cylinder as shown in Figure 8.13, subjected to both bending and torsion. Bending would be caused by the weight of the vehicle, particularly when coming down after driving over a bump and the torsion from cornering. The weight of the vehicle will cause stresses to mount in the tube and will also cause it to deflect. The torsion will likewise result in shear stresses and will cause the tube to twist.

Assuming an even weight distribution, the maximum bending stress σ (N.mm^{-2}) will be given by the formula:

$$\sigma = \frac{wL^4 r_0}{8I} \qquad (8.12)$$

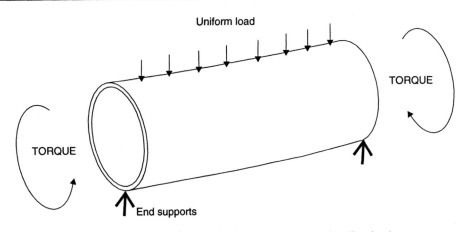

Figure 8.13 A hollow cylinder under torsion and bending loads

where w is the uniform weight/length (N.mm^{-1}), L is the length (mm), r_0 is the radius (mm) and I is the second moment of area (mm^4). I will be given by:

$$I = \pi \frac{r_0^4 - r_i^4}{4} \tag{8.13}$$

And the maximum deflection δ (mm) in the middle of the beam will be given by

$$\delta = \frac{5wL^4}{384EI} \tag{8.14}$$

where E is Young's modulus (N.mm^{-2}). Similarly, the shear stress in the cylinder wall q (N.mm^{-2}) is given by:

$$q = \frac{Tr_0}{J} \tag{8.15}$$

where T is the applied torque (N.mm) and J is the polar second moment of area (mm^4), which will be given by:

$$J = \pi \frac{r_0^4 - r_1^4}{2} \tag{8.16}$$

The angle of twist θ (radians) is given by:

$$\theta = \frac{TL}{GJ} \tag{8.17}$$

where G is the rigidity modulus (N.mm^{-2}).

Certain clear conclusions can be drawn from these equations. To minimise stress due to both bending and torsion both I and J must be kept as large as possible. For a given mass of material, the further it is spread from the centre of the tube, the larger will be both I and J, thus reducing stresses, deflection and twist.

For example, consider a solid cylinder of 200 mm diameter. I and J will be $\pi(100)^4/4 = 25\,000\,000\pi$ and $\pi(100)^4/2 = 50\,000\,000\pi$ mm^4, respectively. The same material can be spread around the circumference of a tube of 1000 mm diameter 10.1 mm thick. This would have values for I and J of $1.23 \times 10^9 \pi$ and $2.45 \times 10^9 \pi$, respectively, an increase of 49 times. The deflections and twists will also be much less for the tube, less by a factor of nearly 50 in fact, as can be seen from equations (8.14) and (8.17).

This remains true until the material buckles, which can be predicted by modern finite element packages. Buckling can be minimised by using two layers of the material, with foam in the middle effectively creating a sandwich, hence sandwich materials. Alternatively two thin sheets of aluminium can be joined by a thin aluminium honeycomb; both of these techniques are widely used in the aircraft industry.

To keep both deflection due to bending and twist due to torque as low as possible it is necessary to use materials which are as rigid as possible, i.e. having high E and G values in addition to optimising the design to keep I and J as large as possible.

Due consideration must be given to material rigidity as well as strength. For example an infinitely strong rubber would be useless as it would deform and twist far too much. Similarly a rigid but weak material would be useless.

Steel, being relatively cheap, as well as rigid, is a traditional choice for manufacturing car bodies and chassis, but it is not necessarily a good choice for electric vehicles. Firstly it has a low strength to weight ratio, resulting in a relatively heavy structure. Secondly the manufacturing cost is low when mass produced, but relatively expensive for small number production, which may be the initial option for electric vehicles.

Materials such as aluminium, and modern composites have much better strength to weight ratios than steels, and are both widely used in the aircraft and racing car industries. A list of some potential materials is shown in Table 8.1 (Kemp 2002).

It can be seen in the table that carbon fibre has the best strength to mass (σ/ρ) as well as the best rigidity to mass (E/ρ), nearly 6 times that of the other materials, which interestingly have almost identical rigidity/mass. This accounts for its widespread use in the aerospace and racing industries. A carbon fibre formula 1 chassis/body can have a mass as low as 35 kg. GRP has the next best strength to mass ratio, 3.5 times that of aluminium which is next.

Before a decision is made on what the appropriate body and chassis materials are, the behavior of a car in a crash must be considered. In a crash situation a car body and/or chassis will absorb energy. If the car is designed so that the energy is absorbed in a controlled manner the forces on the driver and passengers can be minimised. It is therefore normal to design cars with energy-absorbing crumple zones. There is national and international legislation to define a crash situation that cannot be ignored. In the late 1960s one large motor manufacturer had to strip out a brand new production line as the cars they were producing did not comply with crash legislation.

Both metals and composites absorb energy on impact, metals through plastic deformation and composites through fragmenting. The behaviour of metals in a crash can now be predicted accurately using large finite element packages, whereas it is much harder to predict the behaviour of composites. This means that if a metal structure is used the

Design Considerations

Table 8.1 Comparison of material properties

Material	Density ρ (kg/m^3)	Fracture stress σ (MPa)	Young's Modulus E (GPa)	Strength to mass σ/ρ	Rigidity to mass E/ρ
Mild steel	7850	465	207	0.059	0.026
Stainless steel, FSM 1	7855	980	185	0.125	0.024
Aluminium alloy (DTD 5050B)	2810	500	71	0.178	0.025
Magnesium alloy (AX 31) (DTD 742)	1.780	185	45	0.104	0.025
Carbon fibre reinforced plastic, 58% unidirectional fibres by volume in epoxy resin	1500	1050	189	0.70	0.126
Glass reinforced plastic (GRP), 80% uniaxial glass by weight in polyester resin	2000	1240	48.2	0.62	0.024

car can be designed to deform in the optimum manner to meet legislation; this prediction would be much harder with composites.

Both carbon fibre and aluminium are considerably more expensive than steel. However, by using these materials not only is the car lighter, but for a given range a considerable amount of expensive batteries are saved, which must be accounted for in the overall costing of the vehicle.

8.6.4 Designing for stability

As well as being rigid and crash-resistant, it is clearly important that a vehicle design should also be stable. For maximum stability, wheels should be located at the vehicle extremities and the centre of gravity should be kept as low as possible. This is one area where the weight of the batteries can be beneficial, as they can be laid along the bottom of the vehicle, making it extremely stable. During one visit to look at an electric van manufacturer the author was challenged to try to turn it over while driving round roundabouts. Perhaps regrettably, he declined the offer, but it did give an indication of the manufacturer's confidence in the stability of their product. The Duke of Edinburgh drove the same model of vehicle for a while; it is not known if he received the same challenge!

8.6.5 Suspension for electric vehicles

Suspension has the purpose of keeping all of the wheels evenly on the ground, reducing the effects of bumps and ensuring passenger comfort. Suspension on an electric vehicle should, from the energy efficiency viewpoint, be fairly hard. As with tyres pumped to a high pressure, the energy loss is reduced but the ride tends to be less comfortable.

Other than this, the suspension design for electric vehicles will not be different to that for regular vehicles of similar size.

8.6.6 Examples of chassis used in modern battery and hybrid electric vehicles

Some electric vehicles are simply adapted from an internal combustion engine vehicle and use an existing vehicle chassis/body. This has obvious advantages in as much as the whole vehicle is available and simply has to be adapted, which is obviously a cheaper option than designing a whole new vehicle. Although these vehicles often have an adequate range and performance, better results are obtained if the body chassis is purpose built.

Many of the more recent electric vehicles use aluminium for the main structure despite the lower strength/mass of aluminium compared to carbon fibre composites. The vehicle panels are often made from composites.

Some examples are shown in Figures 8.14 and 8.15. The first shows the Twike, a simple elegant design using a tubular aluminium chassis. (The complete vehicle is shown in Figure 8.12.) The aluminium body of the Honda Insight, together with some views of the whole vehicle, is shown in Figure 8.15. The vehicle panels are often made from composites. Note that front crumple zones are a feature of both designs.

8.6.7 Chassis used in modern fuel cell electric vehicles

Chassis bodies for fuel cell vehicles need to house the hydrogen fuel tanks or a hydrogen generator, the fuel cells, the motor and radiators for getting rid of excess heat. Fuel cells can be used in conventional vehicle chassis units, and some examples have been seen

Figure 8.14 Twike chassis

Figure 8.15 Aluminium body from the Honda Insight (Reproduced by kind permission of the Honda Motor Co. Ltd.)

Figure 8.16 General motors Hy-wire chassis (Reproduced by kind permission of General Motors Corp.)

in earlier chapters, for example Figures 1.14 and 1.15. However, General Motors have taken the view that a modern power source required a totally new approach to the design. Their fuel cell vehicle the Hy-wire uses a 'skateboard' chassis illustrated in Figure 8.16.

This chassis unit is based on a simple aluminium ladder frame with front and rear crush zones. The chassis contains hydrogen tanks, drive-by-wire system controls for the steering, cabin heaters, radiators for dispensing with excess heat from the fuel cells and the air management system. The electric motors are built into the wheels. The whole unit is elegant and compact and allows a range of bodies to be attached to the chassis unit, the steering and controls being connected electrically. This allows the chassis to be used as the basis for a range of vehicles, from family saloons to sports cars.

8.7 General Issues in Design

8.7.1 Design specifications

Before anyone sits down to design anything, including an electric vehicle they should write a design specification outlining precisely what they want to achieve. There are books devoted to the subject of writing specifications but it is worth briefly looking at the implications.

For example, is the vehicle required for high speed motorway driving, or is it simply for delivering people or loads about town at low speeds? This fact alone will lead to great differences in the shape of the vehicle. It was seen in the section considering aerodynamics that the vehicle which is to be used for motorway driving needs to be as aerodynamic as possible, but for the low speed delivery van the aerodynamics are of much less importance.

Likewise, although any electric vehicle needs to be protected against corrosion, the environment in which the vehicle is likely to be used needs defining. A vehicle to be used in airport buildings clearly requires much less corrosion protection than one to be used on a seaside pier and constantly subjected to salt water spray. Obviously where vehicles may be used in different environments, the worst case must be allowed for.

The main areas which need specifying for an electric vehicle are range, speed, acceleration, type of use (e.g. passenger commuter car or around town delivery van), performance up hill, legal requirements, target cost (both production and sales). Other parameters that need specifying include life, maintenance requirements, environment, emissions (in the case of a hybrid), aesthetics and comfort.

The design specification must be written bearing in mind technical facts. A battery electric car with a range of 350 miles and a mass of 500 kg and costing £1000 is clearly impossible using today's and foreseeable future technology.

8.7.2 Software in the use of electric vehicle design

Much of the conventional theory as presented in this book and elsewhere is satisfactory for giving first-order calculations and initial systems studies. It is an important initial stage in the design process of electric vehicles to carry out an initial study to check on the likely performance and range of your vehicle. The guidance given in the previous chapter, and in the Appendices, can be used for such analysis.

However, it is usually necessary to use more sophisticated software to more accurately predict the performance of the vehicle. Finite element packages have already been mentioned, and these will give accurate predictions of strength, rigidity and precisely how the body/chassis deforms under load, the dynamics of the body/chassis, and how and where it will vibrate, as well as an accurate prediction (within 1%) of how the body/chassis will crumple in a crash. Likewise the aerodynamic behaviour of the vehicle can be predicted reasonably accurately using computational fluid dynamics (CFD) analysis packages. The actual car will be designed using powerful computer aided design (CAD) programs, and the car will be manufactured using computer aided manufacturing (CAM). Normally large integrated packages containing all of these and using common data from the CAD files are used. Moulds and press tools for body work panels, for example, will be machined from the CAD data that has defined their shape. These will previously have been analysed for airflow using CFD and for strength, rigidity, natural vibration and behaviour in a crash using FE.

It is normal that products such as vehicles going into production will be designed by whole teams of engineers, industrial designers and analysts. Despite this, initial pilot studies for electric vehicles, prototypes and specials can still be designed by individuals or small groups of engineers and designers, and the approach to the design outlined in this chapter will help them in this task.

9

Design of Ancillary Systems

9.1 Introduction

Although the design of major components as discussed in the previous chapter is critical, the design and choice of ancillaries such as the heating and cooling system are also very important.

An important issue in ordinary IC engine vehicles is the ever-rising amount of electrical power required to drive the auxiliary systems. Indeed, this problem is very likely to cause a gradual moving over from 14 to 42 V electrical systems. The average power taken by the electrical systems on even a very ordinary car can be as much as 2 kW. Clearly, this is a particularly important in battery electric vehicles, and most of the problems addressed in this chapter are a particular concern to this type of electric vehicle. The aim will always be to use systems of the lowest possible electrical power.

The heating or cooling system of a car or bus is obviously a major consumer of energy. However, other systems such as steering, and even the choice of wing mirrors and tyres are also important.

9.2 Heating and Cooling Systems

There is little point in producing the ultimate energy-efficient electric vehicle, light, aerodynamic and with high motor and transmission efficiencies, and then waste precious energy by passing current directly through a resistance to heat the vehicle. With IC engine vehicles, copious waste heat will quickly warm the vehicle, although starting off on a cold morning may be unpleasant. For fuel cell vehicles or hybrids with internal combustion engines waste heat is also available, but with battery powered electric vehicles there is little waste heat and where heating is required this must be supplied from a suitable source. Of course, heating does not need to be supplied for electric vehicles such as bikes and golf buggies. Vehicle cooling is often needed in hot climates and this can also absorb considerable energy.

Batteries have a low specific energy and are expensive. It is better to store heat energy by using the specific heat, or latent heat of materials. As an example, one kilogram of water housed in a suitable insulated container and raised through 70°C above ambient contains 293 kJ or 81 Wh of heat. At 81 Wh.kg^{-1} this is a considerably better specific energy than both lead acid and NiCAD batteries.[1] Early night storage heaters used the same principle for storing heat, but they used brick rather than water. More modern night storage heaters use the latent heat of fusion of materials such as wax, which gives an even higher specific energy than that obtained by heating water. Basically the wax is melted and kept in an insulated container. The heat can be drawn from the wax when required. A variation on this theme could be successfully used for storing heat in a vehicle. The heater could be recharged at the same time that the batteries were topped up and heat could be taken off as required. This is the basis of the RHP[2] climate control system of Groupe Enerstat Inc. of Canada. For commuter vehicles this method of heating using thermal stores does have an advantage. A consequence would be that on cold days the vehicle would be warm as soon as the driver gets in, which would be a boon for short journeys.

A similar technique could be used for storing 'cool'. For example ice could be created at night and the latent heat of fusion released when required. The latent heat of fusion of ice/water is 92.7 Wh.kg^{-1} and a further 17.3 Wh.kg^{-1} can be obtained from heating the water to 15°C, giving a total specific energy of 110 Wh.kg^{-1}.

Both of these systems are relatively simple and are worth remembering as methods of heating and cooling electric vehicles. Schematics of both systems are shown in Figures 9.1 and 9.2.

Fuel burning heaters can be used to provide warmth. Such heaters have been used in battery vehicles used by the US postal service. It was said that such heaters could only

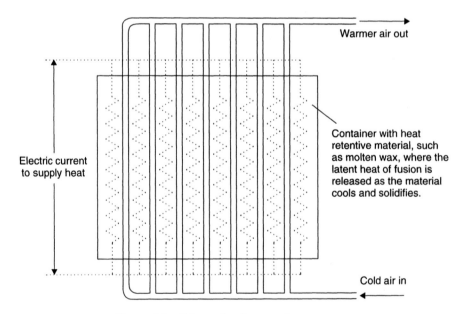

Figure 9.1 Schematic of storage heater system

---

[1] Note that this means the ultra-low technology hot water bottle has a higher specific energy than most types of modern battery.

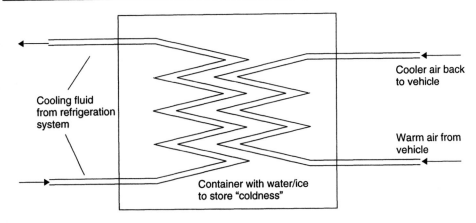

Figure 9.2 Schematic of cooling system

be controlled by opening the doors and letting the heat out, and the result was that the vehicles ended up using almost as much fuel as the diesel powered vehicles they replaced! For all sorts of reasons, this option must be considered a last resort. For one thing, the vehicle can no longer be classified as 'zero emission'.

Another way of heating and cooling an electric vehicle is to use a heat pump, as was done on the GM EV1. A heat pump is a device that actually provides more heat energy than the shaft or electrical energy put in. A schematic of a heat pump is shown in Figure 9.3. As its name implies, a heat pump pumps heat from one location to another.

When heating an electric vehicle, the heat pump would take heat from the outside air and pump the heat to the car interior. A reversible heat pump takes heat from the car and pumps it to the outside air, cooling the car. Heat pumps have a coefficient of performance which is typically 3 or more, in other words for every kilowatt of electrical input, 3 kW or more of heat will be pumped into or taken from the car. Refrigerators and

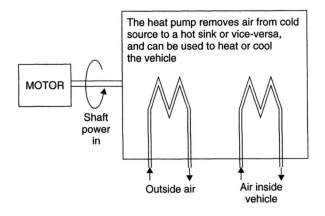

Figure 9.3 Schematic of a heat pump

air conditioning units are examples of heat pumps. When air conditioning both heats and cools it is a reversible heat pump.

The design of heat pumps is complicated, however there are many air conditioning systems available which can be used as a basis. Firms who specialise in heat pumps and air conditioning should be consulted if you need a heat pump for an electric car. Heat pumps are also a feature of the RHP^2 system alluded to earlier.

Car heating is clearly necessary, particularly in cold climates, but there is always some debate as to whether cooling is needed. Normally on a hot day in the UK the windows are opened. However, this will affect the aerodynamics considerably, increasing the drag and shortening the range. It would make better sense to install air conditioning in all electric vehicles where long range is required.

Most building designers know that the best way to heat a building is to keep the heat in and the best way to cool the building is to keep heat out. They also will tell you that the heating needs to come on half an hour before you get up. However, with cars they are either freezing cold, or when in strong sunlight get so hot that pets and small children who cannot free themselves can easily die from the extreme temperatures. Before a heating/cooling system is installed it is worth thinking about the best method of keeping the heat in and out hence minimising both the size of system and also minimising the amount of power used.

Certainly insulating material can be placed around the vehicle. Some modern insulating materials are very thin and would add little to the vehicle mass. Structural foam sandwich materials discussed in Chapter 8 would have good insulating properties and these materials would therefore serve two purposes.

The big problem with overheating stems from the fact that most cars are highly efficient passive solar heating systems; most would make good greenhouses. There is a large glazed area, allowing sun into the vehicle where it is absorbed by the interior as heat. This heat cannot be radiated, convected and conducted away at the same rate and so the vehicle temperature rises, often quite considerably. The sun will also hit the vehicle roof and be conducted through the roof, but this can be cut down by insulating the roof as discussed earlier. This problem can be considerably reduced by using glass covered with a selective coating, which considerably cuts down the amount of solar radiation that enters the vehicle.

There are some other ways in which vehicle heating and cooling can be aided. When at rest in sunshine in hot climates, air can be drawn through the vehicle bringing the internal temperature nearer to the outside air temperature. If the vehicle is being charged from external sources the fan can be powered from this. One neat design is to incorporate a solar panel in the roof to power the fan. It is also worth cooling or heating the vehicle half an hour or so before it is used, either by using a remote control or a time switch. Where a heat pump system is used the electricity would come from the charge point.

9.3 Design of the Controls

Traditional mechanical controls can be used in a traditional way, i.e. steering wheel, and floor mounted accelerator and brake (there is no need for a clutch with an electric

Design of Ancillary Systems

vehicle). However some modern vehicles, such as the GM Hy-wire (shown in Figure 8.16) use more sophisticated modern systems using 'drive by wire'. This is a system which has come from the aircraft industry where it is know as 'fly by wire'. In this system the controls are effectively movement transducers that convert movement into an electrical signal. The electrical signal is normally fed into an electronic controller, or a computer, which in turn controls servos on the brakes, steering, throttle, etc.

There are various configurations for the controls, and they can be configured perfectly normally: steering wheel, accelerator and brake pedals, and the type of 'gear lever' normally found on an automatic transmission car. However, with an all-electric car it is possible to break out of this standard, and use different systems such as the stick controller illustrated in Figure 9.4. Normally a four quadrant electronic controller is used to control the motor, the first quadrant providing forward power and the second providing regenerative braking (the other two quadrants are used in reverse). Because one electronic controller is used for both acceleration and normal braking it is easier to use one lever to accelerate and brake the vehicle. The mechanical brakes could then be added to the extreme lever position. Some manufacturers have experimented with a one stick control, which incorporates both the steering, accelerator and brakes. A stick control fitted to an experimental vehicle is shown in Figure 9.5.

A stick controller can also be used where it is coupled mechanically to the brakes and steering. Normally it is servo-assisted. This system has the advantage that some mechanical control is kept in the event of a breakdown of the power-assisted servos. 'Drive by wire' normally brings out fears of what happens in the event of a failure, but it should be borne in mind that 'fly by wire' has been used successfully on aircraft for years.

An advantage of more modern types of control system that are appropriate for electric vehicles is that computer systems could be used to over-ride user commands. For example, the motor power of a vehicle is normally fixed to allow it to ascend steep hills at reasonable speed, but this power can be used to provide excessive acceleration and speed which wastes energy and reduces range. It is quite practical to provide electronic speed control

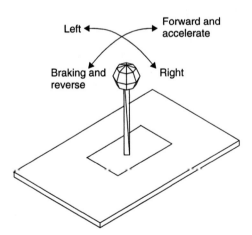

Figure 9.4 Stick controller

Figure 9.5 Stick controller fitted to a vehicle

so that excessive speeds and acceleration are avoided. This could also be linked to satellite navigation systems to ensure that the car never exceeded statutory speed restrictions, increasing road safety as well as maximising vehicle range.

With electric vehicles, where the aim is always to save energy, onboard computers coupled to satellite navigation systems can be even more advantageous. The latter can keep journeys to a minimum by giving precise directions to any destination. They could also direct drivers to nearby charging points.

Design of Ancillary Systems

9.4 Power Steering

Power steering is now standard on many cars, particularly heavier vehicles, which electric cars normally are. With internal combustion engine vehicles it is conventional to use a hydraulic system, the hydraulic pump being powered mechanically from the engine. With electric cars where there is an electrical power source it is easier and more efficient to use electrically powered power steering.

The Honda Insight, for example, uses a variable-assist rack and pinion electric power steering (EPS) system instead of a hydraulic power steering system. A typical hydraulic power steering system is continually placing a small load on the engine, even when no steering assist is required. Because the EPS system only needs to draw electric power when steering assist is required, no extra energy is needed when cruising, improving fuel efficiency.

Electric power steering is mechanically simpler than a hydraulic system, meaning that it should be more reliable. The EPS system is also designed to provide good road feel and responsiveness.

The system's compactness and simplicity offer more design freedom in terms of placement within the chassis. The steering rack, electric drive and forged-aluminium tie rods are all mounted high on the bulkhead, and steer the wheels via steering links on each front suspension strut. This location was chosen in order to achieve a more compact engine compartment, while improving safety.

9.5 Choice of Tyres

The importance of low rolling resistance was highlighted in Chapter 8. Low rolling resistance tyres such as the Michelin Proxima™ RR as used on the GMEV1 have a very low rolling resistance and it is worthwhile to use low energy tyres such as these. There is no compromise in handling and safety stemming from the use of energy-efficient tyres. Hybrid IC/electric vehicles are also normally fitted with such tyres.

Low energy tyres are normally inflated to fairly high pressure, typically 3.5 bar, and this means the ride of the car may be slightly less comfortable, equivalent to a harder suspension. The Proxima™ RR has a special sealant under the tread area that automatically seals small tread punctures. This avoids the need for a spare wheel, which represents a saving in weight, cost and space, all-important parameters in electric vehicle design.

Low energy tyres such as the Proxima™ RR for example are designed to be inherently quiet. This is in itself important as electric vehicles are normally introduced to save environmental pollution and noise pollution is unpleasant. Obviously, it is important for designers to discuss their needs with tyre manufacturers.

9.6 Wing Mirrors, Aerials and Luggage Racks

It is obviously illogical to spend endless time and effort perfecting the aerodynamics of vehicles and then to stick wing mirrors, aerials and luggage racks out of their sides. This immediately increases the aerodynamic drag coefficient, which in turn reduces range.

Modern video systems can be used to replace wing mirrors. Small video cameras are placed at critical spots and relayed to a screen where the driver's mirror is traditionally located. This system has the added advantage of giving better all round visibility. The screen can be split to give information from all round the car at a glance, which would be very useful for city driving where electric vehicles are liable to be used. This system is used on the GM Hy-wire experimental electric car shown in Figure 8.16. The rear view screen is placed in the middle of the steering device.

Aerials can be incorporated on one of the rear windows to avoid external protrusions.

Luggage racks are a more difficult subject, as they may sometimes be needed. Their use will considerably reduce the range of rechargeable battery vehicles. The more aerodynamic the vehicle, the larger will be the percentage reduction in range. It may be better to design battery vehicles so that they do not have the option of any luggage rack or external fitting.

9.7 Electric Vehicle Recharging and Refuelling Systems

Clearly there is no use in introducing electric vehicles without introducing recharging systems for battery vehicles and refueling systems for fuel cell vehicles. The topic of battery charging was covered in Section 2.8 in the chapter on batteries.

In places such as California, and parts of France and Switzerland, where there has been active encouragement of battery electric vehicles, recharging points have been located around cities. Since battery electric cars are usually used for short journeys, of a fairly predictable kind, or at least within a limited region, users will know where charging points are located. Should rechargeable electric vehicles become more widespread, more thought would be needed as to how and where charging points would be situated, and making this information widely known. How the electricity would be paid for would then become more of an issue. In addition, where necessary, suitable electric supply lines would need to be provided and appropriate generating equipment installed.

Plug-in chargers traditionally used conventional transformers containing both primary and secondary windings. More modern plug-in chargers do not need to use transformers. Alternating current is rectified to direct current and this is used to charge a large capacitor. Power electronics are used to switch the current to the capacitor on and off, thus maintaining the DC voltage within narrow bands. (Such 'chopper' circuits are explained in Section 6.2.2.) Transformers contain iron cores and are heavy; eliminating them results in a considerably lighter charging unit. This opens up the way for small onboard chargers, so that battery vehicles can simply be recharged from the mains if no external chargers are available. The majority of electric vehicles carry an onboard charger, though this will usually recharge the batteries at a rather slower speed than is possible with more sophisticated offboard systems.

The problem of battery chargers is one that fuel cell and hybrid electric vehicles do not have at all. However, the problem of supplying fuel to fuel cells is no less complex, and so we devoted the whole of Chapter 5 to this problem. On the other hand, the great majority of hybrid vehicles use the IC engine to recharge the battery, and so simply fill up with gasoline or diesel.

10

Electric Vehicles and the Environment

10.1 Introduction

Mankind is becoming increasingly concerned about the damage it is causing to the environment, and electric vehicles are perceived to play a part in redressing the balance. It is therefore important that the environmental impact of electric vehicles is thoroughly understood.

Ultimately electric vehicles *may* be of substantial benefit, reducing harmful emissions. There is considerable misunderstanding at present as to precisely why electric vehicles can be of benefit, and the extent of that benefit. Firstly, it must be remembered that energy has to come from somewhere, normally power stations; it does not just appear. A key part of the consideration of the environmental impact of vehicles is the so-called 'well to wheels' analysis, where the pollution of all parts of the energy cycle in the use of a vehicle is considered, not just the vehicle itself.

A second point to be borne in mind is that internal combustion engine vehicles can be run entirely from sustainable fuels, as the Brazilian programme of using ethanol made from sugar cane has proved. Internal combustion engines could also be made to run with virtually zero emissions, burning hydrogen for example and thus giving an exhaust gas of (almost) just water and air. Perhaps fortunately, it is becoming easier and more efficient to use fuel cells, and electricity for charging batteries can be derived from renewable sources.

A third aspect is how the availability of electric vehicles could move people towards more environmentally responsible modes of transport. For example, if electric bicycles worked well, and were widely available, could this persuade some people to abandon their private cars, which generate considerable pollution whatever their power source?

10.2 Vehicle Pollution: the Effects

Before we look at solving environmental problems it is worth pausing to look at precisely what environmental problems are caused by vehicles.

Electric Vehicle Technology Explained James Larminie and John Lowry
© 2003 John Wiley & Sons, Ltd ISBN: 0-470-85163-5

There are two main problems caused by conventional vehicles. Firstly they ruin the immediate environment with noise and pollutants. Secondly they burn irreplaceable fossil fuels producing carbon dioxide which is a major cause of global warming and climate change.

You do not need to be a scientist or engineer to understand that motor cars spoil the immediate environment. You simply need to walk along a busy street or sit at a roadside café. The motor vehicle has emerged over a century and we simply accept it as a fact of life. Normally when people who live in the country come to a big city they find both the noise and the fumes quite unacceptable.

The health hazards associated with motor vehicle exhausts are particularly worrying. If you place a stationary diesel engine with the exhaust near a wall, the wall very quickly turns black with what can loosely be described as soot. Again you do not need to be a medical scientist to realise the effect that this might have on your lungs. You would need to smoke a lot of cigarettes to get the same level of deposit, and we all know the health effects of tobacco smoke. Although you do not have to stand behind diesel exhausts, you are bound to inhale a fair amount walking along a busy street and crossing the road, which often involves passing directly through vehicle exhaust.

Accepted health problems associated with car exhausts makes depressing reading and one has to wonder why society keeps quite happily emitting these substances.

The major internal combustion engine pollutants include carbon dioxide, carbon monoxide, nitrous oxides, volatile organic compounds (VOCs), particulate matter, and sulphur dioxide.

Carbon monoxide inhibits the ability of the blood to carry oxygen, and in particular is dangerous to smokers and people with heart disease. It can also cause permanent damage to the nervous system.

Nitrous oxides (NOx) exacerbate asthma, affect the lungs and increase the susceptibility of young children and the elderly to respiratory infections. In the presence of VOCs and sunlight NOx reacts to produce ground level ozone. This in turn irritates the eyes, damages the lungs and causes respiratory problems. NOx contributes to the formation of acid rain whose acidity kills plants and fish. Benzene, a known carcinogen, is an example of a toxic VOC found in vehicle exhaust.

Particulate matter (PM) causes lung problems including shortage of breath, worsens cardiovascular disease, damages lung tissue and causes cancer. Ultra-fine PM makes its way past the upper airway and penetrates the deepest tissue of the lungs and thence to the blood stream. At concentrations above 5 micrograms per cubic metre particulate matter presents a significant cancer risk. Many PMs are recognised as toxicants and carcinogens, as well as hazards to the reproductive and endocrine systems.

New discoveries on the risks of cancer from exhaust fumes continue to emerge. Researchers in Japan have apparently isolated a compound called 3-nitrobenzanthone which is a highly potent mutagen.

Clearly this is cause for alarm. It must also be remembered that new research is constantly emerging and the overall picture may well be extremely grim. Certainly there have been large rises in asthma, many allergies and cancers that may well be linked to exhaust fumes.

Electric Vehicles and the Environment

The effect of carbon dioxide on the planet is another cause for alarm. The greenhouse effect of carbon dioxide is now well known. Basically, some of the short-wave radiation from the sun is absorbed by the earth and then re-emitted at a longer wavelength. This is absorbed by carbon dioxide and other gases and then re-emitted, the downward radiation warming the surface of the earth. The atmospheric concentration of carbon dioxide has increased by about 25% over the past 100 years.

Although a warmer earth may sound appealing to those living in cold climates, there are side effects which could prove absolutely devastating. Firstly the earth relies on a reasonably set weather pattern for growing food. A change of climate in the grain growing belt of America, for example, could itself have serious consequences on food supply. Secondly the 'warm up' is melting the polar ice caps and this could cause permanent flooding in low-lying areas. Bearing in mind that many major cities, London, New York, Barcelona, San Francisco, Perth (Australia) and scores of others, are built on the coast, this could have very serious repercussions throughout the world.

One significant problem with internal combustion engine vehicles in slow traffic is that fuel consumption rises very dramatically as vehicles crawl along at slow speeds and pollution gets considerably worse. This is illustrated in Figure 10.1. With electric vehicles there will be a small decrease in efficiency of the electric motor when used at low speeds but the efficiency of batteries such as lead acid increases resulting in a fairly steady efficiency across the speed range. In cities such as London and Tokyo the average speeds are normally less than 15 kph and in rush hour are considerably less.

The simplest way of eliminating these problems from town and city streets is to enforce zero emission vehicles into the towns and cities by legislation or other means.

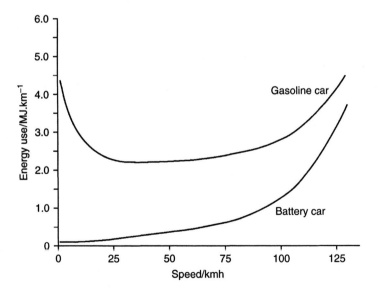

Figure 10.1 Indicative energy use for IC engine and battery powered cars. Obviously the precise figures vary very greatly with size and design of vehicle. These figures are *not* the whole well-to-wheel energy figures, but just the tank-to-wheel or battery-to-wheel figures

Conventional internal combustion vehicles ruin the environment in their vicinity particularly in towns and cities.

The simplest way of creating zero emission vehicles is to adopt electric vehicles, or at least hybrid vehicles which solely run on electricity when in the town and city environment. However, the total pollution impact of vehicles and their energy use cannot be ignored, and this we consider in the next section.

10.3 Vehicles Pollution: a Quantitative Analysis

Energy for electric vehicles clearly has to come from somewhere. If battery electric vehicles are widely introduced the vast majority will have to be charged from the mains grid, where at present most electricity production comes from burning fossil fuels. At present sustainable sources of energy currently provide less than 10% of the energy used in the grid, so most of the electricity used for charging electric vehicles would be obtained from burning fossil fuels, including coal, gas and oil, at the power stations.

Conversion efficiency (energy at power station outlet/calorific value of fuel) for producing electricity from fossil fuels at modern power stations is typically about 45%, much higher than motor-car engines. However, this has to be transmitted to the consumer and the average transmission efficiency, including transmission through the low voltage local networks, is around 90%. This means that the actual efficiency of converting the chemical energy of fuel at the power station to energy at consumers' electrical socket outlets is typically about 41%. This then has to be converted to energy delivered via the vehicle wheels.

For an efficient electric vehicle the efficiency of converting electrical energy supplied to the vehicle on charging to energy at the wheels will be around 50–60%. This means that the overall efficiency of converting fuel at the power station to wheel energy for electric vehicles is around 20%. The overall efficiency of internal combustion engine vehicles (energy delivered via the wheels/fuel energy) under normal driving is typically 12–18%, a very similar figure to that of electric vehicles. The result is not really surprising; basically one combustion engine, a vehicle bound diesel or petrol, with a transmission system, is being replaced with another located at the power station.

However, as discussed earlier, fuel efficiency of internal combustion engine vehicles at low speeds in heavy traffic gets considerably worse; see Figure 10.1. Under these conditions electric vehicles undoubtedly produce considerably less carbon dioxide, which in itself makes a strong environmental case for their use, particularly in towns and cities. Also, it should be borne in mind that harmful emissions from burning fuel at a power station can and should be very carefully monitored and controlled and where possible eradicated.

This study of the pollution caused by the use of a vehicle is usually called a 'well-to-wheel' analysis. In this context a coal mine, for example, is also a 'well'. There are several stages in the process, and these will depend greatly on the vehicle. In any case, every stage of the process will involve pollution and/or the consumption of energy. See Figure 10.2.

Several different fuel cycle analyses or well-to-wheel studies have been published, including ETSU (1996) and Hart and Bauen (1998). There is also a good summary of

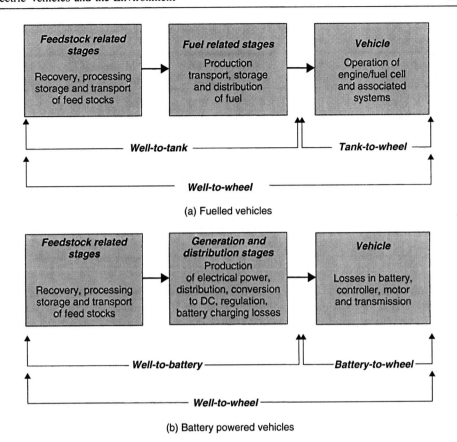

Figure 10.2 Energy transfers in fuelled and battery powered vehicles used in the analysis of Section 10.3

this research by Bauen and Hart in Hoogers (2003). Some of the results of their analysis are shown in Table 10.1.

For this study the emissions of the gasoline and diesel IC engine cars were taken as those for the EURO III standard. The energy consumption, and hence carbon dioxide production, figures were taken from the average consumption of the all British cars in 1998. The average engine efficiency was taken as 15%.

The car fuelled on compressed natural gas (CNG) was assumed to have the same energy use as the gasoline vehicle, but with a 10% better efficiency, i.e. 16.5% efficient.

The hydrogen fuelled IC engine vehicle was assumed to be supplied with cryogenic liquid hydrogen, as outlined in Chapter 5. This is the method most commonly used among those companies, such as BMW, that have made the most progress with this technology. It is explained in Chapter 5 that the liquefaction of hydrogen is a very energy-intensive process, and this has a significant effect on the energy use results.

For the hydrogen fuelled fuel cell electric vehicle, the hydrogen was assumed to be generated using medium-scale steam reforming plant based at refilling stations, with natural

Table 10.1 Emissions and energy use figures for different types of car, being mainly a summary of data calculated by Bauen and Hart (2003)

Vehicle type	NOx, g.km^{-1}	SOx g.km^{-1}	CO g.km^{-1}	PM g.km^{-1}	CO$_2$ g.km^{-1}	Energy MJ.km^{-1}
Gasoline ICE car	0.26	0.20	2.3	0.01	209	3.16
Diesel ICE car	0.57	0.13	0.65	0.05	154	2.36
CNG ICE car	0.10	0.01	0.05	<0.0001	158	2.74
Hydrogen ICE car	0.11	0.03	0.04	0.0001	220	4.44
Gasoline fuelled hybrid	0.182	0.14	1.61	0.007	146.3	2.212
MeOH fuel cell car	0.04	0.006	0.014	0.0015	130	2.63
Hydrogen fuel cell car	0.04	0.01	0.02	<0.0001	87.6	1.77
Battery car, British electricity	0.54	0.74	0.09	0.05	104	1.98
Battery car, CCGT electricity	0.17	0.06	0.08	0.0001	88.1	1.71

gas as the feedstock. The hydrogen would then be stored at high pressure (~300 bar) onboard the fuel cell vehicle, as outlined in Chapter 5. The consensus view of studies in the USA, and summarised by Ogden (2002), is that this is the most economic and least environmentally damaging method of supplying hydrogen to vehicles.

The second fuel cell vehicle uses hydrogen derived from an onboard methanol reformer, as described in Chapter 5. The performance is assumed to be broadly similar to the units described there, and realistic figures from the methanol industry were used for the feedstock and fuel stages (see Figure 10.2).

The hybrid vehicle is rather difficult to include, as there is so much variation in the degree of hybridisation, as was mentioned in Chapter 1. The emissions per kg of petrol consumed are unlikely to be very different from those of ordinary IC engine vehicles; the main benefit is that less fuel will be used, and proportionately less pollution of all types produced. So, for the hybrid vehicle the figures have been set, quite simply, as 70% of those of an ordinary gasoline powered vehicle. For the next 10 years or so, this represents a reasonable figure for the saving that might be expected from such vehicles, though of course this is highly debatable.

It is impossible to find figures for a battery electric vehicle that are not contentious. Following the very thorough study in ETSU (1998), the energy figure of the electric vehicles[1] is assumed to be 0.72 MJ.km^{-1}. The simulation described in Chapter 7 can be used to show that the energy taken from the battery by the GMEV1 electric car driving the SFUDS cycle, with heater and headlights on, is about 0.4 MJ.km^{-1}. So a fair allowance is being made for a rather more harsh driving cycle, which is quite justified. Inefficiencies in the generation, distribution, and charging process mean that the actual energy use is about two and a half times this figure. In their study, Bauen and Hart (2003) proposed two different sources, and these figures have been used. One was the average figure from the current mix of UK electricity generators. This comprises a combination of generator

---

[1] The energy figure is the energy out of the battery, so due allowance is made for losses in the motor, controller, transmission, etc.

types that is probably not very different from that of many western countries. The second vehicle was assumed to be powered by electricity from state-of-the-art combined cycle gas turbine (CCGT) generators supplied with natural gas. Such a hypothetical system makes a good comparison with the hydrogen fuelled fuel cell vehicles, which also use a somewhat idealised supply system. The efficiency of CCGT systems is about 50%, whereas that for the current British system is about 43% on average. The charging process is assumed to be about 85% efficient, which is obtainable when using NiMH batteries.

Not included in this table are three types of vehicle that are considered later in the chapter:

- battery powered vehicles, with batteries charged from renewable electricity generators;
- fuel cell vehicles, powered by hydrogen made from biomass or water electrolysed by renewable electricity;
- IC engine vehicles powered by biofuels such as ethanol.

These have zero net carbon dioxide emission, and very low production of other pollutants, though the last is the worst in this regard. However, their overall energy use and efficiency cannot sensibly be estimated at the moment, as there is not enough experience with this technology to make sensible comparisons.

The figures of Table 10.1 are shown in the graphs of Figures 10.3, 10.4 and 10.5. To a certain extent these charts speak for themselves, but several interesting points can be made:

1. Figure 10.3 shows that all the electric vehicles are noticeably better than all the IC engine vehicles.
2. The source of the electrical power for the battery powered vehicles is very important. Figure 10.5 shows that a battery powered vehicle using the current British electricity generation mix is not significantly better as regard pollutants than other vehicle types.
3. The hydrogen fuelled fuel cell vehicle comes out very well from the study.
4. The hydrogen powered IC engine appears very poor from the point of view of energy use. This is because an inefficient fuel processing system is combined with an inefficient engine. However, the non-carbon dioxide pollutants are very low.
5. The hybrid vehicle also comes out very well from the study. If hybrid technology were combined with a diesel engine, then even better performance might be expected.

10.4 Vehicle Pollution in Context

The extent and importance of the pollution from vehicles is sometimes questioned, especially by certain motorist lobbying organisations. Table 10.2 shows figures for Britain regarding energy usage over the years 1990 to 2000. They show clearly how the energy used on the road has gradually risen, and indeed the energy used for transport has even risen slightly as a proportion of all energy used.

The amount of carbon dioxide produced is more-or-less directly proportional to the mass of petroleum used, though it should be remembered that the table is not really

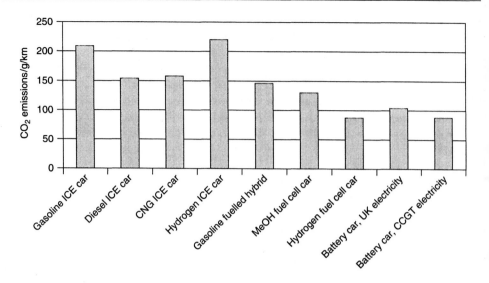

Figure 10.3 Data from Table 10.1 showing the carbon dioxide emissions (well-to-wheel) from different categories of vehicle

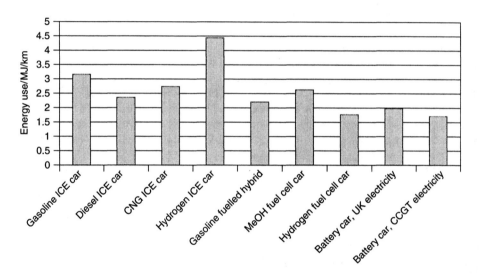

Figure 10.4 Data from Table 10.1 showing the energy use (well-to-wheel) arising from different categories of vehicle

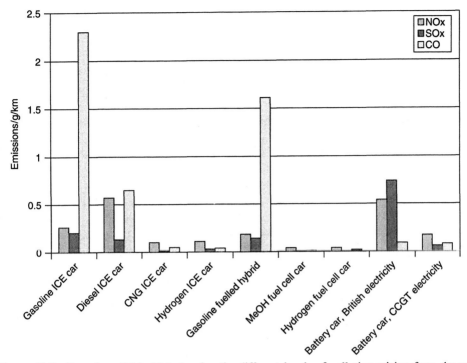

Figure 10.5 Data from Table 10.1 showing the different levels of pollution arising from the use of different types of car

Table 10.2 Data for energy use by the transport sector in the UK. Except for the last column, the figures are the two year averages. The figures are in millions of tonnes of petroleum equivalent energy

	1990/1	1992/3	1994/5	1996/7	1998/9	2000
Road	38.68	39.43	39.48	41.02	41.21	41.07
Railways	0.68	0.70	0.65	0.58	0.51	0.48
Water transport	1.36	1.36	1.22	1.28	1.13	1.04
Aviation	7.1	7.66	8.28	9.12	10.63	11.86
All energy used by transport	48.31	49.78	50.25	52.67	54.2	55.20
All energy used by final users	149.55	151.92	151.48	155.86	157.54	160.06
Energy used by transport as a percentage of all energy used by final users	32.3	32.8	33.2	33.8	34.4	34.5

talking about tonnes of petroleum, but an energy equivalent, and a move to more natural gas and less coal will help a little. Road transport represents about 25% of the energy used, and thus about 25% of the green house gas production. It will be much the same in other western countries. The conclusion is that a more efficient transport system can make a very substantial contribution to the reduction in greenhouse gas emission.

In terms of other pollutants, personal transport probably has an even greater impact. Especially in our cities, a very high proportion of such air pollution comes from IC engines.

10.5 Alternative and Sustainable Energy Used via the Grid

There is no reason at all why alternative sustainable forms of energy such as solar, wind, hydro, tidal, wave, biomass, waste and energy taken from under sea currents should not be used to provide energy to electric vehicles via the grid. In some ways this is a much better option than powering electric vehicles directly from small wind generators and solar panels, where the energy is wasted when the vehicle is not directly charging. With the grid the energy can be transferred from one use to another. When electricity is not being supplied to one consumer it will be supplied to another. When there is ample electricity from alternative sources it will be used, fossil-fuelled generators being switched off to save fuel. When there is less alternative energy the fossil-fuelled generators will be used, the customer not experiencing any break in supply during periods of little sun, wind or tide. Also, the grid allows electric vehicles to be supplied with energy from all available energy sources including hydro, tidal, wave and under sea currents.

Although most electricity is currently obtained from burning fossil fuels at power stations, there is a positive effort being made to introduce alternative sustainable forms of energy. These could be used to provide electricity for rechargeable battery cars and also to provide and process fuels for metal air batteries and fuel cells.

Using alternative forms of power to energise electric vehicles would make a real impact on the environment, totally eliminating emissions and pollutants from motor vehicles. Perhaps society should make a concerted effort, not only to introduce electric vehicles, but at the same time to install sufficient sustainable energy to power these vehicles. It is therefore appropriate to examine the various alternatives to see how they can be used.

10.5.1 Solar energy

Solar energy was discussed briefly in Chapter 3 in relation to small scale use with electric vehicles. On a large scale it can be used to supply electricity to the grid or to use it to electrolyse hydrogen for use in fuel cell vehicles. The maximum efficiency for commercial solar cells is currently about 16.5%, which drops to around 14% when encapsulated. Given the current drive to maximise the performance of solar photovoltaic cells this is likely to increase. The maximum power of solar radiation on the earth's surface is $1000\,Wm^{-2}$, but is normally less due to cloud and dust in the atmosphere.

The actual cost per peak kilowatt of solar cells is around $4500, the installed cost will be higher. In order to convert the direct current electricity from the solar panels to AC electricity for the grid, an inverter is required. This will also need to synchronise the frequency with the grid frequency, and deal with a multitude of safety issues relating to this type of equipment.

Solar panels for production of electricity can be mounted on the ground or alternatively on buildings. Interestingly, the amount of solar energy falling on all buildings in the UK is about 1600 TWh per year, as compared to fossil fuel use of about 1500 TWh per year. Although efficiencies of conversion must be accounted for, this is still a substantial amount.

10.5.2 Wind energy

Wind energy, as with solar, is a rapidly developing technology. Growth of wind energy production has averaged at 40% per annum and it is likely to go on expanding at this rate. In the early 1980s the largest commercially available wind turbine was 50 kW. By the end of the 20th century 1.7 MW machines were commercially available. The total wind power installed in Europe is 20 447 MW and the British Isles 655 MW. This equates to around 60 000 000 MWh per annum in Europe and around 2 000 000 MWh in Britain. To produce this amount of energy by burning oil at a power station with an overall efficiency of 0.33 would require 18 million tonnes of oil in Europe and 600 000 tonnes of oil in Britain compared to 41 000 000 tonnes of oil used for road transport in Britain. Wind energy currently available in Britain could possibly provide 1.5% of the energy needed for transport if it were used to charge electric vehicles. Although this is a relatively low figure the UK is only capturing 0.5% of the wind energy available. It would therefore be possible, in theory at least, to provide most if not all of the energy needed for transport by wind energy if required.

Large wind farms can now be found all over the world, many producing up to 40 MW of electrical power. An example large (750 kw) wind turbine is shown in Figure 10.6. This is from the wind farm in El Perello, Spain, 'Parc Eolic de Colladeres'. This park has 54 large turbines giving a total maximum power of 36.63 MW. These typically produce 97.5 million kWh per annum. Again, if this energy were produced in a conventional fossil fuel burning power station of efficiency 33% it would require 30 000 tonnes of oil.

10.5.3 Hydro energy

Hydro energy has been used successfully for several thousand years, initially in the form of water wheels to drive mills. The principle of hydro power is essentially simple, running water used to drive a turbine. In large hydro schemes a valley in a hill or mountain is dammed and a lake formed. Outlet pipes from the dam direct water through a water turbine. The water flow is controlled to give power on demand.

The surprisingly high figure of 6% of world electricity generation is currently obtained from hydro power. In the UK 2% of power is obtained, compared to Canada where the figure is 60%. There is potential to increase the amount of hydro electricity, current thinking encouraging a multiple of smaller hydro schemes.

10.5.4 Tidal energy

Tidal energy on a small scale has been used for centuries on the coasts of Britain and France. Proposals for a major barrage in Britain were published as early as 1849. Probably the best known tidal scheme is the Rance power station in France. In this scheme the tidal energy is captured by damming the estuary and forcing the tidal waters through axial flow

Figure 10.6 750 kW wind turbine from the Parc Eolic de Colladeres, Spain

turbines. The scheme typically produces about 500 GWh of electricity per year; 150 000 tonnes of oil per year would need to be burnt in a conventional power station to produce this amount of electricity.

10.5.5 Biomass energy

Energy can be obtained from plants, which are in effect a form of stored solar energy, the solar radiation being used to form chemical energy. There are two main ways of using biomass. The first is to grow plants such as sugar cane, extract the sugar and turn this into ethanol, which can be used as a fuel for IC engines. This could either be used to run IC engine vehicles directly, as has been done in Brazil for several years, or could be used as a fuel at suitable power stations. The second is to grow fast-growing trees, which are coppiced annually. The wood is then used as fuel in a wood burning power station to generate electricity. Typically 10 to 20 tonnes per hectare of wood fuel can be produced per annum from wood plantations.

10.5.6 Geothermal energy

Geothermal energy is produced by taking heat from underground rocks and running this through a heat engine and generator to produce electricity. Normally water is pumped underground via a pipe and returns to the surface via a second pipe. Provided too much heat is not taken away when the rocks become chilled, this method is sustainable. Such energy is only usable in a few locations, Iceland being a case in point. Interestingly, they are actively working on conversion of this energy to chemical energy in the form of hydrogen, for use in fuel cells.

10.5.7 Nuclear energy

There are two types of nuclear energy, nuclear fission and nuclear fusion. Conventional nuclear power stations use nuclear fission and are currently in service providing electricity. The nuclear lobby point out that nuclear power stations do not give off gaseous pollutants or carbon dioxide. However, there is concern about the nuclear waste and in many countries the nuclear program is gradually being phased out.

Nuclear fusion creates energy by fusing hydrogen into helium. The principle is used in the hydrogen bomb, but the energy release is uncontrollable. Recently nuclear fusion research has been carried out in the JET project at Culham, UK. Although there is a long way to go before nuclear fusion becomes a commercial reality, it still holds out great hope for the future.

10.5.8 Marine current energy

A considerable amount of energy can be captured from undersea currents; certainly this is the opinion of companies working in the field, such as Marine Current Turbines Ltd. The company plan to install up to 300 MW in the next decade. Up to 20 to 30% of the UK's electricity needs could be provided from this source alone. Up to 48 TWh per year could be produced from 106 sites around Europe, the majority being in the UK.

Around 48 TWh of electricity would require the burning or 15 million tonnes of oil (assuming an efficiency of conversion of 33%), which in theory could power one-third of the UK's transport energy requirements. The marine currents are more predictable than wind and solar.

10.5.9 Wave energy

Another method of obtaining sustainable energy is by tapping the power in waves. There are several systems under trial but as yet none has been commercialised. None the less, this is another strong contender for providing energy supply in the future.

10.6 Using Sustainable Energy with Fuelled Vehicles

10.6.1 Fuel cells and renewable energy

Fuel cells are about twice as efficient as IC engines, and as such could in theory halve the amount of energy used in transport. Should fuel cells become sufficiently developed for general use in motor vehicles, they will undoubtedly make a significant impact on the production of carbon dioxide and almost eliminate production of other pollutants, as can be seen from Figures 10.3–10.5.

Hydrogen for fuel cells can either be obtained by reforming conventional fuels such as natural gas or by electrolysing water. This could be done using small local units, with people generating their own local supply of hydrogen for use in their own vehicles. As is explained in Chapter 8 of Larminie and Dicks (2003), it is entirely practical to use electrolysers to generate hydrogen directly at high pressure to pass into a high pressure store without any need for pumping.

The latter method can be carried out using sustainable forms of energy such as those discussed above. One advantage of using hydrogen as a fuel is that its production can be matched to alternative energy availability. For example it can be produced at times of high wind or solar energy availability. It can also be produced by conventional generating plant at off peak times, which enhances the general efficiency of electricity production.

Hydrogen can also be made using renewable biological methods, and from biomass and waste. Such ideas are discussed, for example, in Chapter 8 of Larminie and Dicks (2003).

10.6.2 Use of sustainable energy with conventional IC engine vehicles

It must be remembered when considering the energy impact of electric vehicles that conventional IC engine vehicles can also use sustainable energy, as mentioned before. In Brazil for example vehicles have been fuelled by ethanol, made from sugar cane for nearly 3 decades. About 14 billion litres of ethanol are produced annually from sugar cane grown for this purpose.

Vehicles with modified internal combustion engines could also be run from hydrogen, although the combustion process would still cause some pollutants. As the efficiency of fuel cells is higher than that of IC engines it is likely that fuel cells will eventually become the chosen option.

Bio diesel is another source of sustainable fuel for internal combustion engines. Bio diesel is normally made from the esters of vegetable oils such as palm oil or olive oil.

Efficiencies of IC engine vehicles, together with the quietness and the cleanliness of their exhausts, improve all the time, taking the sting out of some of the criticisms of conventional vehicles. This must be borne in mind when comparing electric vehicles with those with internal combustion engines.

10.7 The Role of Regulations and Law Makers

The question which is bound to be asked is: if society develops suitable electric vehicles using rechargeable batteries or fuel cells, can we use only sustainable energy for transport and totally eliminate exhaust pollutants? The answer is undoubtedly 'yes' particularly in the long term, and if society is prepared to pay for this. However, the power of law and regulation is important in making individual people make choices, in favour of less pollution at a higher price. One of the major roles of society is a 'collective coercion to be good,' and this can be seen very clearly in the case of electric vehicles.

This is best illustrated by the actions of California through its Air Resources Board (CARB). This organisation was a major promoter of the now almost universal catalytic converter on IC engine exhausts. It has had a huge impact on the development of electric vehicles.

The story began in the late 1980s, when the CARB enacted a directive that required that any motor manufacturer selling vehicles in the state would have to ensure that 2% of vehicle sales in 1998, rising to 5% in 2000, would have to be zero-emission vehicles. The California vehicles market is huge, about 1 000 000 per year, so this had massive implications, which the motor manufacturers reacted to with great energy. Two major consequences were the production of high technology vehicles like the General Motors EV1 described in the following chapter, and also major developments in fuel cells. However, despite great efforts, it became clear that the targets were highly unrealistic. Also, developments in hybrid vehicles, and a greater understanding of the total well-to-wheel emissions of battery vehicles, which we have been discussing here, led to constant revision of the regulations.

These revisions came in two forms. The first was a rolling back of the targets in time. The second was to make them much more complex, though this complexity correctly reflects the complexity of the issues. Vehicles that are not fully zero-emission might actually produce fewer emissions in the whole well-to-wheel analysis than a ZEV. So, although it is easy to criticise the CARB for caving in to motor industry pressure, this would not really be fair. The regulations now incorporate a system of credits, which a motor manufacturer can use. So, for example, it can sell two vehicles with credit 0.5 and this counts as 1 fully ZEV vehicle. The latest indications[2] are that manufacturers will not need to produce any fully ZEV vehicles as part of their fleet of sales.

It is very interesting, in the light of the analysis of vehicle pollution given in Section 10.3, to look at the table of credits given for different types of vehicles by the CARB. These can be seen in Table 10.3. The first three vehicles in this table also featured in Table 1.1 of Chapter 1. Several of the others were also in above Table 10.1. Grid hybrids are hybrid vehicles where the battery is somewhat larger, and which can be recharged from the mains grid, as well as the onboard generator.

The total credit is an indication of the perceived 'environmental value' of a vehicle. Many of the figures can be compared with vehicles shown in Figures 10.3–10.5. It can be seen that these credit table seem to have been drawn up with similar data in mind. The only major divergence would appear to be the very high credit being proposed by the CARB for the hydrogen ICE car, which does not appear to be entirely justified. Many readers, especially those in Europe, will also find the absence of any reference to diesel hybrids strange.

Regulations such as these from the CARB will certainly be a very major influence on the future development of electric vehicles. In particular, they will encourage the

---

[2] In April 2003.

Table 10.3 Credit table, as being proposed by the CARB in April 2003, for different types of PZEV, for the years 2005 to 2011 (CARB 2003)

Vehicle type	Zero emission range	Base credit	Zero emission range credit	Advanced components credit	Low fuel cycles emission credit	Total credit
Low voltage HEV	0	0.2	0	0	0	0.2
High voltage HEV	0	0.2	0	0.4	0	0.6
High voltage, high power HEV	0	0.2	0	0.5	0	0.7
CNG ICE car	0	0.2	0	0.2	0.3	0.7
CNG ICE based hybrid (>10 kWe)	0	0.2	0	0.6	0.3	1.1
Hydrogen ICE car	0	0.2	1.5	0.3	0.3	2.3
MeOH fuel cell car	0	0.2	1.5	0.5	0.3	2.5
Grid hybrid with 20 miles electric range	20	0.2	1.25	0.5	0.12	2.1
Grid hybrid with 30 miles electric range	30	0.2	1.4	0.5	0.15	2.3
Grid hybrid with 60 miles electric range	60	0.2	1.82	0.5	0.15	2.7
Hydrogen ICE based hybrid (>100 kWe)	–	0.2	1.5	0.7	0.3	2.7
CNG ICE based hybrid with 20 miles electric range	20	0.2	1.25	0.7	0.3	2.5

development of a wide range of hybrid vehicles, which the data indicates is a very good way of reducing the total well-to-wheels environmental damage of vehicles.

References

Bauen A. and Hart D. (2003) Fuel Cell Fuel Cycles, pp. 12-1 to 12-23 of Hoogers (2003) below.
CARB (2003) Description and Rationale for Staff's Additional Proposed Modifications to the January 10, 2003 ZEV Regulatory Proposal, www.arb.gov.ca March 2003.
ETSU (1996) *Alternative road transport fuels: a preliminary life-cycle study for the UK*, Vol. 2, HMSO, London.
Hart D. and Bauen A. (1998) Further assessment of the environmental characteristics of fuel cells and competing technologies, Report #ETSU F/02/00153/REP, Department of Trade and Industry, UK.
Hoogers G. (Ed.) (2003) *Fuel Cell Technology Handbook*, CRC Press, Boca Raton, FL.
Larminie J. and Dicks A. (2003) *Fuel Cell Systems Explained* 2nd Edn, Wiley, Chichester.
Ogden J.M. (2002) Review of Small Stationary Reformers for Hydrogen Production, IEA Report no. IEA/H2/TR-092/002.

11
Case Studies

11.1 Introduction

At present many exciting developments in electric vehicle technology are taking place. Some of these have advanced sufficiently to be commercially available, whilst others remain for the future.

During the last few years a variety of modern electric vehicles have come on the market for purchase, lease or for prototype trials. In this chapter some state-of-the-art electric vehicles that are available commercially, or which have undergone extensive trials, are discussed. Some of these are outstanding examples, some are simply representative. There are many fine examples not discussed simply because there is insufficient space to include them.

In Chapter 1 we outlined six different types of electric vehicle. However, of these only the first three have made any real commercial headway. These are firstly rechargeable battery vehicles, secondly hybrid vehicles and thirdly electric vehicles which can be refuelled using either fuel cells or metal air batteries.

11.2 Rechargeable Battery Vehicles

Rechargeable battery vehicles can also be divided into several different categories. For example, there are electric bicycles, secondly there are the low speed vehicles (LSVs) which form a class of vehicle in the USA and Canada, with maximum speeds of 40 kph (25 mph), and thirdly conventional road vehicles using rechargeable batteries. In addition there are special purpose delivery vehicles and vehicles such as fork lift trucks. There are also the small four-wheeled carriages used by the infirm which can be ridden on the pavement, and are narrow enough to fit through normal front doors. Other vehicles, such as powered wheel chairs, could also be mentioned.

11.2.1 Electric bicycles

Electric bicycles are probably the most popular type of battery electric vehicle. For example, thousands are used in China. There are many different manufacturers and types, with a very interesting range of power methods: hub motors in the front or back wheels, and drives on the pedal cranks are the most common variations. In most European and North American countries it is becoming a standard regulation that these bikes must be of the 'pedal-assist' type. This means that they cannot be powered by the electric motor alone. If they can be ridden under electric power only, then they count as motorbikes, and attract a host of extra regulations and taxes. However, the regulatory situation is somewhat fluid, varied and changeable, which is something of an impediment to the development of this market.

For example, in the UK 'electric power' only bicycles are still legal (in 2003), but the power must cut out once the cycle reaches a speed of 15 mph. The maximum allowed motor power is 200 W, the maximum weight is 40 kg, and the riders must be aged 14 years or over. However, this law is liable to change in the near future in favour of pedal assist mode only.

These regulations mean that the controllers on electric bikes nearly always include a sensor of some kind. Often there will be a torque sensor, on the pedal crank. This works with the motor controller, and ensures the rider is putting in some effort before allowing the motor to provide any power. Speed sensors will also often be fitted, to cut the electric power once a set speed is reached, 15 mph in the UK.

Although the large cycle manufacturers, such as Giant of Taiwan, are including electric cycles, the market is currently dominated by smaller businesses supplying a fairly local region. For example, in the UK one notable manufacturer is Powabyke Ltd., and one of their cycles is shown in Figure 11.1. Some technical details of their range of

Figure 11.1 Electric bicycle

Table 11.1 The specification for the Powabyke range of electric cycles

Wheels	24 or 26 inch
Frame	17 inch
Motor	200 W Hub Motor
Speed control	Handlebar-mounted twist throttle
Range, pedal assist mode	Claimed 50 km, approximately
Battery	36 V, 14 Ah VRLA battery
Gears (for pedalling)	5, 21 or 24 speed
Battery charge time	8 h
Mode	Pure power, pedal assist or pedal only, switchable
Charger	36 V mains charger
Weight including battery	39 kg
Current cost	About £699.99 depending on specification

standard cycles are given in Table 11.1, though they also produce other types, such as a foldable cycle.

The development of reliable electric bikes could in itself have a significant impact. It may encourage people to use bikes rather than their conventional internal combustion engine vehicles. A survey by the manufacturers showed that the average Powabyke covered 1200 miles per annum, replaced 3 car journeys per week, provided a journey time faster than a bus and cost less than 1.5 pence per mile to run.

Use of this type of vehicle could be increased with encouragement from governments and councils in the form of special cycle tracks. The use of electric bikes would undoubtedly lead to less vehicle traffic and decreased pollution.

11.2.2 Electric mobility aids

One area of application that the demographics of western countries indicate must steadily grow for the foreseeable future is mobility aids for the elderly and infirm. The sort of vehicle we are talking about here is shown in Figure 11.2, a carriage for those who can take a few steps, but need help from technology to retain their independence. There is a wide range in this class, from small three-wheelers to larger vehicles with a claimed range of 40 miles, and tougher wheels, for people who need to get about country lanes. Table 11.2 gives the outline technical details of a middle range machine such as that of Figure 11.2.

11.2.3 Low speed vehicles

Low speed vehicles (LSVs) are an environmentally friendly mode of transport for short trips, commuting and shopping. In the USA, for example, 75% of drivers are believed to drive round trips of less than 40 km per day. In rural areas the lower traffic density would enable these to be used fairly easily. In towns and cities it would be worthwhile for governments and local authorities to ensure that proper lanes for this type of vehicle were made available, and where possible tax incentives are used to encourage their use.

Figure 11.2 Mobility aid for shopping trips and similar journeys near home

Table 11.2 Details of a shopping carriage for the infirm, such as that shown in Figure 11.2

Length	1.19 m
Width	0.6 m
Height	0.91 m
Weight	65 kg, 85 kg with batteries
Batteries	2 of 12 V, 28 Ah (20 hour rate)
Motor power	370 W continuous
Drive	Rear wheel drive, one motor
Maximum speed	6 kph forward, 2 kph reverse
Maximum incline	10°C
Range	30 km (claimed, on level ground)

These vehicles are particularly targeted at fairly active retired people, who still want to get about to see their friends, but who do not travel so far, are not in such a hurry, and value a peaceful neighbourhood. The demographics of most western countries show there is little doubt that the market for this type of vehicle will grow steadily.

An example LSV is shown in Figure 11.3. The specification of a typical four seat vehicle of this type is given in Table 11.3.

Figure 11.3 An example of the low speed vehicle (LSV) type of electric car

Table 11.3 Specification of a typical low speed electric vehicle

Length	3075 mm
Width	1400 mm
Height	1750 mm
Wheelbase	2635 mm
Curb weight	560 kg
Payload	360 kg
Ground clearance	150 mm at curb weight
Minimum turning radius	4080 mm
Range	64 km
Acceleration	0 to 40 kph, 6 s
Hill climbing capability	21%

11.2.4 Battery powered cars and vans

There is a range of vehicles, vans, cars and buses that have been used in recent years. Most have ranges under 100 km, and the best possible range of a rechargeable battery vehicle such as the GM EV1 is 200 km. While this market will no doubt always continue at a small level, there is no evidence of much growth, and indeed recent developments in California, mentioned in the final section, mean that this market may well decline.

An example is the Ford Th!nk shown in Figure 1.6. This is a two-seat car using a steel chassis with thermoplastic body panels. It has a top speed of 90 kph and accelerates to 48 kph in 7 seconds. It uses 240 kg of NiCad batteries and has an overall weight of 940 kg. Despite a very expensive development programme, this car is no longer in production, as Ford see the future more in hybrids and fuel cell cars.

Peugeot have always been very active in electric vehicles, and work with vans as well as cars. Two example vans are the 106 and the Partner. The Partner van is illustrated in Figure 11.4 and the specification of both is given in Table 11.4. There is also a 106 electric car, which has a similar specification to the van.

The Peugeot electric vehicles are examples of conventional vehicles converted to electric vehicles. Their range is doubtless adequate for uses such as city deliveries, but is less than might have been obtained with a purpose-built electric vehicle. This is undoubtedly a trade-off of cost against range.

When designing their EV1 vehicle, as shown in Figure 11.5, General Motors took another philosophy altogether. The GM EV1 is a rechargeable battery powered two-seater car, which represents probably the most advanced electric vehicles using rechargeable

Figure 11.4 Peugeot Partner battery electric van. (Photograph reproduced by kind permission of Peugeot S.A.)

Case Studies

Table 11.4 Details of Peugeot battery electric vans

	106 Van	Partner Van
Motor type	Leroy Somer DC separate excitation 20 kW	Leroy Somer DC separate excitation 28 kW
Max. motor torque	127 Nm from 0–1600 rpm	180 Nm from 0–1550 rpm
Max. motor speed	6500 rpm	6500 rpm
Motor cooling	Forced air ventilation	
Transmission	Front wheel drive, epicyclic reduction to differential	
Steering	Rack and pinion, power assisted	
Suspension, front	Independent with MacPherson type struts	
Suspension, rear	Independent with trailing arms and torsion bars	
Front brakes	Servo-assisted discs	
Rear brakes	Servo-assisted drums	
Tyres	165/70 × 13	165/60 × 14
Maximum speed	90 kph	96 kph
Range (urban)	72 km	64 kph
Restart gradient	22%	
Battery type	NiCad, 100 Ah	
Voltage	120 V	160 V
Charger	3 kW integral with vehicle 240 V, 13 A, AC supply	3 kW integral with vehicle 240 V, 13 A, AC supply
Charging time	7 h	9 hours
Energy consumption	20 kWh/100 km	28 kWh/100 km
Length	3.68 m	4.11 m
Width	1.59 m	1.69 m
Height	1.38 m	1.81 m
Kerb weight	1077 kg	1450 kg
Payload	300 kg	500 kg
Payload volume	0.92 m^3	3 m^3

batteries. It was purposely designed through and through as an electric vehicle and is not a modified IC engine vehicle. It was introduced in 1997, initially using lead acid batteries; more recent versions have used NiMH batteries to give an extended range.

We have already met this car when considering vehicle modelling in Chapter 7. It has the lowest drag coefficient of any production car ($C_d = 0.19$) and uses very low rolling resistance tyres. The vehicle was produced in limited numbers and could only be leased rather than bought. To the dismay of electric vehicle enthusiasts, the EV1 was recently taken off the market. The recent decision of the California legislature to draw back from its requirement on manufacturers to produce fully 'zero emission' vehicles has probably made this decision permanent. The existing examples will no doubt become valuable collectors' items in the years to come, because it is a very interesting car with many exciting technical features. The EV1 is illustrated in Figure 11.5.

Figure 11.5 The groundbreaking General Motors EV1 battery electric car (reproduced by kind permission of General Motors Inc.)

The EV1 is powered by a 102 kW, three-phase AC induction motor and uses a single-speed dual reduction gear set with a ratio of 10.946:1. The battery pack consists of 26 valve-regulated high-capacity lead-acid batteries, each 12 V and 60 Ah. The EV1 can be charged safely in all weather conditions with inductive charging. Using a 220 V charger, charging from 0 to 100% takes from five-and-a-half to six hours. The EV1 with the high-capacity lead-acid pack has an estimated real-world driving range of 50 to 90 miles, depending on terrain, driving habits and temperature (see Section 7.4.2.3). The range with the optional NiMH pack is even greater. Again, depending on terrain, driving habits, temperature and humidity, estimated real-world driving range will vary from 75 to 130 miles.

Braking is accomplished by using a blended combination of front hydraulic disk and rear electrically applied drum brakes and the electric propulsion motor. Regenerative braking is used, extending the vehicle range by partially recharging the batteries.

The vehicle's body weighs 132 kg and is less than 10% of the total vehicle weight, which is 1400 kg, of which the battery weight is nearly 600 kg. The 162 pieces are bonded together into a unit using aerospace adhesive, spot welds and rivets.

The exterior body panels are dent and corrosion-resistant. They are made out of composites and are created using two forming processes known as sheet moulding compound (SMC) and reinforced reaction injection moulding (RRIM).

The EV1 is designed to be highly aerodynamic, saving energy and allowing a lower level of propulsion power that sends the vehicle further. The rear wheels are 9 inches closer together than the front wheels, which allows for a tear-drop body shape that lessens drag, as explained in Section 8.2.2.

The EV1 has an electronically regulated top speed of 80 mph (130 kph). It comes with traction control, cruise control, anti-lock brakes, airbags, power windows, power door locks and power outside mirrors, AM/FM CD/cassette, and tyre inflation monitor system. The vehicle also offers programmable climate control, an electric windshield defogger/de-icer, a rear window defogger and centre-mounted instrumentation.

The EV1 does not require a conventional key to unlock the door. A five-digit personal identification code is entered on the exterior keypad to allow access. No key is needed to start the car. The same five-digit code is entered on the centre console's keypad to activate the car.

The EV1 undoubtedly has as good performance and range as can be achieved economically using commercially available rechargeable batteries. The advanced performance clearly illustrates the benefits of designing the vehicle as an electric car rather than simply converting an existing vehicle. However, even this highly advanced design could not stand up to the market competition. The development in hybrid electric/ICE powered vehicles has probably also been a major contribution to its failure as a profitable product with significant sales.

The range of rechargeable battery vehicles is normally limited to below 150 km and at best 200 km. The use of batteries such as the Zebra battery or lithium chloride batteries could increase range further by 30%. Clearly, for a very much greater range a new generation of batteries is required.

Although the major manufacturers are losing interest in battery vehicles, the new low speed vehicles and electric bikes are making headway, as is the market for vehicles for the elderly and infirm, and it is in this area where future development is likely to take place.

11.3 Hybrid Vehicles

Adding an additional source of power such as an engine generator unit, or even a supply rail, is a simple way to extend the range of rechargeable battery powered vehicles. However, this is not at the moment the way in which electric vehicle development is moving. The trend among manufacturers is to develop vehicles in which the engine and motor are used in conjunction to optimise the vehicle fuel economy. Two examples are the Honda Insight and the Toyota Prius, which have really made an impact on the world of car design, and brought electric cars that people can easily use on to the market. These, and a steadily increasing number of alternatives from almost all the major motor manufacturers, can be purchased now at very reasonable prices.

11.3.1 The Honda Insight

The Honda Insight, illustrated in Figure 8.14, is a hybrid vehicle combining a conventional gasoline-driven engine with an additional motor driven by a battery. The engine and motor can both be used to propel the vehicle. The Insight employs a system Honda calls the Integrated Motor Assist (IMA) system. The Insight is not recharged from the mains in the same way as a conventional electric vehicle. Instead its benefits derive from using the internal combustion engine in conjunction with the battery and motor/generator system to maximise energy efficiency and to minimise fuel consumption. When there is surplus power available from the engine it is used to recharge the batteries from the motor/generator. The motor/generator is also used to slow the vehicle, and thus recover the kinetic energy into the battery. When the brake pedal is pressed lightly, the Insight's electric motor operates in regeneration mode, and the car begins to slow just as it would with normal brakes. Once the brake pedal is pressed further, the normal brakes come into play, slowing the car down even more.

In heavy traffic the car is driven from the batteries via the motor/generator only. Hence it can be classified as a partial zero emission vehicle (PZEV). It is a parallel hybrid, as outlined in Figure 1.10. The electrical motor/generator has a maximum power of 10 kW, and is about 6 cm thick, between the engine and the gearbox, and directly connected to the crankshaft.

The body chassis is designed for low weight whilst at the same time meeting crash test requirements. All structural components and most body panels are extruded or die-cast aluminium, while front fenders and rear fender skirts are a recyclable ABS/nylon composite. The car's independent front suspension uses lightweight, forged-aluminium suspension arms and aluminium front suspension knuckles. The braking system's front calipers are also aluminium alloy, as are the rear brake drums.

The rear suspension is a highly compact twist-beam design that sits completely below the Insight's flat cargo floor, along with the lightweight plastic-resin 40 L (10.6 US gallon) fuel tank and gas-pressurised rear shock absorbers. The rear suspension is also designed to help absorb the energy of a rear impact.

At the heart of the Insight's IMA system is a compact, 1 litre, three-cylinder gasoline IC engine. The engine uses lean-burn technology, low-friction design features and lightweight materials such as aluminium, magnesium and special plastics, in combination with a new lean-burn-compatible NOX catalyst, to achieve high efficiency and low emissions. The specification for the Insight is given in Table 11.5.

Table 11.5 Details of the Honda Insight hybrid electric/IC engine car, taken from Honda sales information[1]

Engine size	1 litre, 3 cylinder VTEC
Electrical energy storage	NiMH battery, 144 V, 6.5 Ah
Fuel consumption	26/29 km.L^{-1} city/highway[2] (EPA estimates)
Power	50 kW at 5700 rpm without assist, 56 kW at 5700 rpm with assist
Torque	91 Nm at 4500 rpm without assist, 113 Nm at 1500 rpm with assist
Electric motor	Permanent magnet, brushless DC type, 60 mm thick
Electric motor power	10 kW max. at 300 rpm
Front suspension	MacPherson strut with aluminium forged knuckle, aluminium lower arm
Rear suspension	Twist-beam and trailing arms
Brakes	Four-wheel ABS, front disc, rear drum
Launch dates	1999 (Japan, USA), 2000 (UK, Canada), 2001 (Australia)
Length	3.95 m
Width	1.35 m
Weight	834 kg
Maximum speed	166 kph
Range	Over 1100 km, 40 l gasoline tank

[1] Taken from the Honda website at www.honda.com.
[2] Multiply km.L^{-1} figure by 2.35 to get a result in miles per US gallon, or by 2.82 for miles per UK gallon.

11.3.2 The Toyota Prius

At about the same time as the Insight was launched, Toyota also launched it Prius vehicle. This is also a hybrid gasoline IC engine/electric hybrid. It has less good fuel consumption figures, but has more luggage space and five seats. It has enjoyed considerable sales success, and has really put this type of vehicle into the public eye. This has been helped by the appeal of the car to a number of celebrities.[3] The car is shown in Chapter 1, Figure 1.11.

The car is powered by a 16 valve four cylinder engine, which uses variable valve timing. The engine displacement is 1.5 l with a bore of 75 mm and stroke of 84.7 mm. The engine also incorporates an aluminium double overhead cam (DOHC) and a multi-point electronic fuel injection. This system allows the engine to maintain a high level of fuel efficiency, so that controlled quantities of fuel are used on each combustion cycle.

The vehicle uses an electronic ignition system, which incorporates the Toyota Direct Ignition system (TDI). With regards to performance, the engine can produce a power of 52 kW at 4500 rpm and a maximum torque of 111 Nm at 4200 rpm. Further performance details are given in Table 11.6

The hybrid element is provided by an electric motor and a separate generator, so unlike the Honda Insight it has two electrical machines, and is not a pure parallel

Table 11.6 Technical details of the Toyota Prius[4]

IC engine size	1.5 litre, 4 cylinder, 16 valve
ICE Power	52.2 kW at 4200 rpm
ICE Torque	111 N.m at 4200 rpm
Electrical motor power	33 kW
Electric motor torque	350 Nm at 0–400 rpm
Electrical energy storage	NiMH battery, 288 V, 6.5 Ah
Hybrid system net power	73 kW
Fuel consumption	22/19 km.L^{-1} city/highway (EPA estimates)
Transmission	ECCVT, electronically controlled continuously variable transmission
Suspension	Independent MacPherson strut stabiliser bar and torsion beam with stabiliser bar
Steering	Rack and pinion with electro-hydraulic assist
Brakes	Front disc, rear drum, with ABS
Length	4.31 m
Width	1.69 m
Height	1.46 m
Wheelbase	2.55 m
Weight	1254 kg
Gasoline tank capacity	44.7 l, 11.8 US gallons
Tyres	P175/65R14 low rolling resistance

---

[3] Leonado de Caprio, Cameron Diaz and Meryl Streep are among those said to be enthusiastic owners of these vehicles in 2003.

hybrid. The motor type is a permanent magnet, as with the Honda, which is able to produce a power output of 33 kW. This motor is able to sustain a maximum torque of 350 Nm at 0–400 rpm, which is enough to move the car at slow speeds. The battery required is a nickel-metal hydride (Ni-MH) system, consisting of 228 cells, giving 6.5 Ah at 288 V.

The transmission system is an electronically controlled continuously variable transmission (ECCVT) which give a better performance over the range of gears. The transmission incorporates a fairly complex system of planetary gears, called a power splitter, which directs power between the IC engine, the electric motor, the generator and the wheels, in all directions! A display on the dashboard gives a continuous indication of where energy is going. For example, when accelerating hard energy will be going from both the IC engine and the electric motor to the wheels. When at an easy steady speed, energy will go from the engine to the wheels, and also from the engine through the generator back to the battery. When slowing, energy will go from the wheels, through the generator and to the battery. This display is fascinating, indeed perhaps a little too interesting to watch.

The suspension uses an independent MacPherson strut with stabiliser bar at the front of the vehicle and a torsion beam with stabiliser bar at the rear. The steering column uses a rack and pinion system with electro-hydraulic power-assist and is able to achieve a turning circle of 30.8 ft.

Power-assisted ventilated front discs and rear drums with standard Anti-lock Brake System (ABS) and regenerative braking enable the vehicle to stop in a manner which prevents skidding, even when braking heavily into a corner. Traction is provided by standard P175/65R14 low rolling resistance tyres on aluminium alloy wheels.

The car fits five people, and has a good-sized luggage space not noticeably reduced in size because of the battery, which is stored under the rear passenger seat.

11.4 Fuel Cell Powered Bus

In Chapters 1 and 4 we introduced various fuel cell powered vehicles. However, for a case study we will present the type of fuel cell vehicle that is most likely to make a commercial impact in the medium term, the fuel cell powered bus.

There are several important reasons why fuel cell systems are even more promising in city bus applications than for other types of vehicle. The three most important are as follows.

1. Fuel cells are expensive, so it does not make sense to buy one, and then leave it inactive and out-of-use for most of the day and night, which is the state of most cars. Buses, on the other hand, are in use for many hours each day.
2. The supply of hydrogen for fuel cell vehicles is such a difficult problem that we devoted a whole chapter of this book to it. Buses, on the other hand, refuel in one place, so only one refilling point needs to be supplied.

---

[4] Taken from Toyota sales information at www.toyota.com.

Case Studies

3. The advantages of zero emissions at the point of use are particularly important for city vehicles, which is exactly what this type of bus is, all its life.

It is not surprising then that buses feature strongly among the most exciting demonstration fuel cell vehicles. A modern design was shown in Figure 1.16. Figure 11.6 shows an earlier example that was used between 1998 and 2000 in Vancouver and Chicago. Some high altitude trials were also carried out in Mexico City. The layout of the fuel cell engine is shown in Figure 11.7, and had a maximum power of 260 kW. Ballard made a good deal of data on this system available, and further information can be deduced by calculations from the given data, as presented in Chapter 11 of Larminie and Dicks (2003).

Referring to Figure 11.7, the system consisted of two fuel cell units (5), each consisting of 10 stacks, each of about 40 cells in series. So the total number of cells was about 800, giving a voltage of about 750 V. In use the voltage fell to about 450 V at maximum power. The voltage was stabilised to between 650 and 750 V using a DC/DC converter, as outlined in Section 6.2. There were several step-down DC/DC converters to provide lower voltages for the various subsystems, such as the controller (2), and to charge a 12 V battery used when starting the system. These voltage conversion circuits are unit (1) in Figure 11.7.

The fuel cell system is water cooled, with a 'radiator' and electrically operated fan (3). This could dispose of heat at the rate of 380 kW. As was explained in Section 4.6, fuel cell systems need to get rid of more heat than IC engines of equivalent power. This cooling system also removes heat from the motor (4) and the power electronics (1).

Figure 11.6 Fuel cell powered bus, 1998 model (reproduced by kind permission of Ballard Power Systems.)

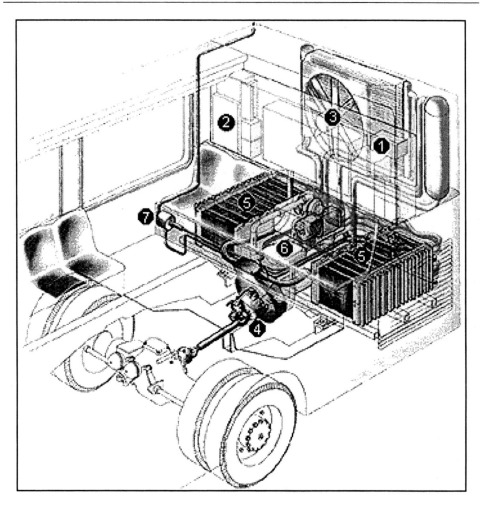

Figure 11.7 Fuel cell engine for buses based on 260 kW fuel cell (diagram reproduced by kind permission of Ballard Power Systems.)

An ion exchange filter was used to keep the water pure, and prevent it from becoming an electrical conductor. Clearly then, no anti-freeze could be used, and so the system had to be kept from freezing, which was done using a heater connected to the mains when not in use. This is one important improvement seen on the more modern fuel cell buses such as that of Figure 1.16. All the losses are dealt with by this cooling system, so we note that 380 kW seems an appropriate value for a 260 kW fuel cell, and suggests an efficiency for the fuel cell of about 41% at maximum power, from the calculation:

$$\eta = \frac{\text{output}}{\text{output} + \text{losses}} = \frac{260}{260 + 380} = 0.41 = 41\%$$

The efficiency at lower powers will be a little higher than this. The motor (4) is rated as 160 kW continuous, which means that for short periods it could operate at about 200 kW. The motor was normally of the BLDC type explained in Section 6.3.2. There is evidence (Spiegel *et al.* 1999) that some models of this bus used induction motors, which illustrates very well what we said in Chapter 6 about the type of motor used being relatively unimportant. Induction motors are rugged and lower in cost, BLDCs are slightly more efficient and compact. Dynamic braking was used to reduce wear on the friction brakes, but not regenerative braking (see Section 6.1.7). The motor is coupled to the forward running drive shaft via a 2.42:1 fixed gear, and to the rear axle via a differential, which will have a gear ratio of about 5:1, as in Figure 8.6 and Section 8.4.

If the fuel cell output is 260 kW, and the maximum motor power is about 200 kW, where does the remaining 60 kW go? The major 'parasitic' power loss is the air compressor (6), which is needed as the fuel cell operates at up to about 3 bar (absolute). As was explained in Chapter 4, this increase in pressure brings performance benefits, but takes energy. Even using a turbine, which extracted energy from the exhaust gas, the electrical power required to drive the compressor will have been about 47 kW. The other major power losses will have been in the power electronics equipment, about 13 kW assuming 95% efficiency, and for the fan to drive the cooling system, probably about 10 kW. These three loads explain the 'missing' 60 kW.

This bus used compressed hydrogen tanks stored on the roof of the bus. These posed no greater safety problems than those present in a normal diesel fuelled bus. Any rupture of the tank, and the hydrogen would rapidly dissipate upwards and out of harm's way. The pressure of the tanks was about 250 bar when full, which was reduced to the same pressure as the air supply, about 3 bar. Usually when the pressure of a gas is reduced greatly, there is usually a cooling effect, but this does not happen with hydrogen. The hydrogen behaves very differently from an ideal gas, and the so-called Joule-Thompson effect comes in to play and there is actually a very modest temperature rise of about $7°C$ in the pressure regulation system.

Much was learnt from the generally successful trials with these buses over several years. This information has been incorporated into the new design of buses, such as those of Figure 1.16, and those from other non-Ballard companies such as the MAN bus of Figure 4.2. People are more likely to take a ride in a fuel cell bus before they go for a drive in a fuel cell car.

11.5 Conclusion

The future of electric vehicles, both in the short and the long term is very exciting. Firstly, there have been considerable developments in technology, which now allow advances in electric vehicle design to be made. Secondly, there are growing environmental concerns which are pressing society to find alternatives to IC engines alone as a source of power for vehicles. Environmental concerns encompass worries about carbon dioxide emissions and the effect of exhaust gas emissions on health. Thirdly, in the largest market for personal transport, the USA, there is an increasing realisation that fuel economy is important, for security reasons as well as environmental concerns. The Californian car market alone is about 1 000 000 units per year, and the rules of this state will continue to give a 'technology push' to developments in this area, as they have done so strongly up to now.

There have been three main areas where substantial developments in electric vehicles are currently occurring. The first is small rechargeable battery vehicles, secondly hybrid vehicles, and thirdly fuel cell vehicles.

There has been a proliferation of small-scale commuter vehicles, bikes, delivery vehicles, mobility aids and lightweight cars that use rechargeable batteries. These have a limited range and are normally used as second vehicles. As such they encourage people to use this form of transport for short journeys and hence are reducing the use of conventional vehicles. Although there have been developments in rechargeable battery cars, such as the GM EV1, the major manufacturers seem to be fighting shy of this area, preferring to develop hybrid and fuel cell vehicles. However, there are many types of vehicle other than cars, and battery electric vehicles of different types (such as bicycles) will become more and more common.

Hybrid vehicles have developed very rapidly in the last few years. The Toyota Prius has been a particular success, at least as a technology demonstration if not commercially, and nearly all major motor manufacturers are developing products in this area. This is being encouraged by regulatory changes in California, and tax breaks throughout the USA. Current practice is to use engine and battery in conjunction to maximise fuel economy, rather than to charge the vehicle from an external electric charging point. However, this may change, and grid connectable hybrids will certainly appear in the medium term. High fuel economy is obtained from these hybrids. Car makers will definitely be pushing forward with hybrids, even though they may have somewhat abandoned battery powered cars. Table 11.7 was produced by Ford ahead of the launch of their hybrid electric SUV car in 2003. It makes clear their view that with a hybrid you get many of the fuel economy and environmental benefits of a battery car, but without having to pay too much in up

Table 11.7 A comparison chart adapted from that produced by Ford,[5] which summarises why they think hybrid IC/electric vehicles are better than 100% battery cars, at least for the medium term

	Conventional	Hybrid electric	100% Battery electric
Total range	350 miles	450–550 miles	100 miles
Electric range	Nil	Nil	100 miles
Gasoline range	350 miles	450–550 miles	Nil
Fuel economy	Base	30–50% better than base	100 mpg equivalent
R-fuelling	Fill-up, 5 min	Fill-up, 5 min, less often	Several hours recharge
Environmental impact	base	SULEV, 90% better than 2003 standard	ZEV, but see Chapter 10, total emissions *not* zero
Performance	four-cylinder	Like a V6	Very poor
Price premium	Base	Not determined, but usually a few $1000	$10 000

---

[5] Found at www.ford.com.

front cost, and getting a vehicle that actually performs better than the standard IC engine only car.

Most major vehicle manufacturers are also currently making developments in fuel cell vehicles. Clearly they see this as an area where electric cars could be produced that would compete with conventional IC vehicles in terms of range, flexibility and cost. Fuel cell cars are further away from commercialisation than hybrids, but fuel cell powered buses are closer to the market.

In addition to developing electric vehicles close attention needs to be given to the infrastructure needed to supply power for electric vehicles. Although small electric commuter vehicles use household electric sockets at present, and current commercially available hybrid vehicles solely use gasoline or diesel, future fuel cell vehicles are likely to need sources of hydrogen. More widespread use of rechargeable battery vehicles will require charging points to be installed.

The issue of energy sources also needs to be addressed. Introduction of electric vehicles undoubtedly cleans up the immediate environment where the vehicle is being used. However, in the case of rechargeable vehicles the emission of carbon dioxide is simply being transferred to fossil fuel burning power stations. Introduction of more alternative energy power stations such as solar, wind and hydro to match the introduction of electric vehicles would ensure real zero emission transport.

With current technical developments in the energy sources for electric vehicles, coupled to the desire for less environmentally damaging transport, the future for electric vehicles looks extremely promising.

Rapid development of small rechargeable vehicles, electric hybrids and fuel cell electric vehicles is likely to continue over the next two decades. At the same time the infrastructure for powering electric vehicles will develop. It is hoped that more emphasis will be placed on the provision of clean sustainable energy systems to provide electric power for rechargeable vehicles, and to produce hydrogen for fuel cells.

References

Larminie J. and Dicks A. *Fuel Cell Systems Explained*, 2nd Edn. Wiley, Chichester.
Spiegel R.J., Gilchrist T. and House D.E. (1999) Fuel cell bus operation at high altitude. *Proceedings of the Institution of Mechanical Engineers*, Vol. 213, part A, pp. 57–68.

Appendices: MATLAB® Examples

These appendices list, with suitable explanations, the MATLAB® programs or 'script files' that are used in Chapter 7, where electric vehicle modelling is considered. In most cases the simulations can be performed nearly as easily with programs such as EXCEL®. However, in some cases MATLAB® is noticeably better. It is also certainly much easier to explain what has been done, and to relate it to the underlying mathematics with MATLAB®.

All the example programs here will work with the student edition of MATLAB®, which can be obtained at very reasonable cost.

Appendix 1: Performance Simulation of the GM EV1

In Section 7.3.3 we gave the MATLAB® script file for simulating the acceleration from a standing start of an electric scooter. In Section 7.3.4 we turned our attention to the groundbreaking GM EV1, and developed some equations for its performance. The script file for modelling the acceleration of that vehicle is quite similar to that for the scooter, and is given below.

```
% GMEV1   *********************
% Simulates the WOT test of the GM EV1 electric car.
t=linspace(0,15,151);   % 0 to 15 seconds, in 0.1 sec. steps
v=zeros(1,151);         % 151 readings of velocity
dT=0.1;                 % 0.1 second time step
% In this case there are three phases to the acceleration, as
% explained in the text.
for n=1:150
    if v(n)< 19.8
       % Equation 7.21
       v(n+1) = v(n) + dT*(3.11 + (0.000137*(v(n)^2)));
    elseif v(n) > 35.8
       % Controller stops any more speed increase
       v(n+1) = v(n);
    else
```

```
        % Equation 7.22
        v(n+1) = v(n) + dT * ((62.1/v(n)) - 0.046 - (0.000137*(v(n)^2)));
    end;
end;
v=v.*3.6;    %Multiply all v values by 3.6 to
             %convert to kph
plot(t,v);
xlabel('Time/seconds');
ylabel('velocity/kph');
title('Full power (WOT) acceleration of GM EV1 electric car');
```

The output of this script file can be seen in Figure 7.6.

Appendix 2: Importing and Creating Driving Cycles

An important part of range modelling is the use of standard driving cycles, as explained in Section 7.4.1. When using MATLAB® these driving cycles must be set up as one-dimensional arrays which contain successive values of velocity. There are many ways of doing this, including methods involving directly reading files. However, the easiest way to do the task is to simply set up a matrix by cutting and pasting in the data as text into a file, as follows.

Let us imagine a very simple driving cycle involving increasing the speed at $1\,\text{ms}^{-1}$ each second, up to $3\,\text{ms}^{-1}$, and then slowing down again. This would be set up as a MATLAB® script file like this:

```
V = [ 0
1
2
3
2
1
0];
```

This would be saved under a name such as SIMPLE.m and would be called from the simulation program.

For real cycles the values would be obtained from a file from the WWW, and then cut and pasted, making a file like the one above, except that it would be several hundred lines long. Each line would just have one number. Such files should be saved under suitable names, and will be called early in the simulation program as shown in the examples that follow. Some of the most common, including those discussed in Chapter 7, have been made available as M-files on the website associated with this book.[1]

Because each cycle will have a different number of values, the simulation program will need to find how many numbers there are, so that it cycles through the correct number of points. This is done using the MATLAB® function length().

Once the velocity values are loaded into an array, care must be taken with the units, as all simulation work must be carried out in ms^{-1}. Thus, in some of the programs that follow all the values in the matrix V are divided by 3.6, to convert from km.h^{-1} to ms^{-1}.

---

[1] www.wileyeurope.com/electricvehicles.

The first three lines of a range simulation will often contain the lines:

```
sfuds;  % Get the velocity values.
N=length(V);  % Find out how many readings
V=V./3.6;
```

As we saw in Section 7.4.1, not all driving cycles are a simple string of values, but are calculated from a series of values of accelerations, followed by so many seconds at such a speed, and so on. This sort of cycle can reasonably easily be created, and an example is given below. It creates an ECE-47 driving cycles for the scooter shown in Figure 7.3. The velocity/time graph produced by this program (or script file) was shown in Figure 7.12.

```
% ECE_47       *****************
% This file is for constructing the velocity
% profile of the ECE-47 cycle for an
% electric scooter.

% The cycles last 110 seconds, so we set up a
% one dimensional array for 111 readings.
V=zeros(1,111);
% The first phase is a 50 second acceleration phase.
% We use exactly the same program as in Section 7.3.2,
% except that the graph drawing elements are removed,
% and the conversion to kph.
Scoota;
% "Scoota" finds values of velocity every 0.1 seconds,
% so we need to decimate these readings.
for n=1:51
    V(n)=vel(((n-1)*10)+1);
end;
%The velocity is then reduced to 5.56 m/sec over the
%next 15 seconds.
decel=(V(51)-5.56)/15;
for n=52:65
    V(n)=V(n-1) - decel;
end
%This velocity is then maintained for 35 seconds
for n=66:101
    V(n)=5.56;
end;
%The scooter is then stopped over 8 seconds
decel=5.56/8;
for n=102:108
    V(n)=V(n-1)-decel;
end;
V(109)=0;
% The zero speed is then held for a further 2 seconds.
V(110)=0;
V(111)=0;
% *****************
```

In order to produce diagrams such as Figure 7.12 plot commands are added at the end of the file. However, when doing range and other simulations, as outlined in Section 7.4, these should not be used.

In all this work, and the examples that follow, it is important to note that with MATLAB® variables are normally 'global'. This means that a variable or array created in one file can be used by another.

Appendix 3: Simulating One Cycle

The simulation of the range of a vehicle involves the continuous running of driving cycles or schedules until there is no more energy left. The script file below is for the simulation of just one cycle. It is saved under the name one_cycle.m. It is called by the range simulation programs that follow. Broadly, it follows the method outlined in Section 7.4.2, and the flowchart of Figure 7.14.

This file requires the following:

- An array of velocity values V must have been created, corresponding to the driving cycle, as outlined in the previous section.
- The value of N must have been found, as also explained in the previous section.
- Two MATLAB® functions, open_circuit_voltage_LA and open_circuit_voltage_NC must have been created. These functions have been outlined and explained in Chapter 2.
- All the variables such as mass, area, Cd, etc., must have been created by the MATLAB® file that uses this file. Rather than listing them again here, refer to either of the programs in the two following sections.

```
% *******************************
%   ONE CYCLE
% This script file performs one cycle, of any
% drive cycle of N points with any vehicle and
% for lead acid or NiCad batteries.
% All the appropriate variables must be set
% by the calling program.
% *******************************

for C=2:N
    accel=V(C) - V(C-1);
    Fad = 0.5 * 1.25 * area * Cd * V(C)^2;  % Equ. 7.2
    Fhc = 0;                % Eq. 7.3, assume flat
    Fla = 1.05 * mass * accel;
        % The mass is increased modestly to compensate for
        % the fact that we have excluded the moment of inertia
    Pte = (Frr + Fad + Fhc + Fla)*V(C);  %Equ 7.9 & 7.23
    omega = Gratio * V(C);
    if omega == 0 % Stationary
        Pte=0;
        Pmot_in=0; % No power into motor
        Torque=0;
        eff_mot=0.5; % Dummy value, to make sure not zero.
    elseif omega > 0 % Moving
```

```matlab
      if Pte < 0
         Pte = Regen_ratio * Pte; % Reduce the power if
      end;          % braking, as not all will be by the motor.
      % We now calculate the output power of the motor,
      % Which is different from that at the wheels, because
      % of transmission losses.
      if Pte>=0
         Pmot_out=Pte/G_eff; % Motor power> shaft power
      elseif Pte<0
         Pmot_out=Pte * G_eff; % Motor power diminished
      end;           % if engine braking.

      Torque=Pmot_out/omega; % Basic equation, P=T * omega
       if Torque>0 % Now use equation 7.23
          eff_mot=(Torque*omega)/((Torque*omega)+((Torque^2)*kc)+
          (omega*ki)+((omega^3)*kw)+ConL);
        elseif Torque<0
         eff_mot=(-Torque*omega)/((-Torque*omega) +
         ((Torque^2)*kc)+(omega*ki)+((omega^3)*kw)+ConL);
       end;
       if Pmot_out > = 0
          Pmot_in = Pmot_out/eff_mot;  % Equ 7.23
       elseif Pmot_out < 0
          Pmot_in = Pmot_out * eff_mot;
       end;
end;

 Pbat = Pmot_in + Pac;  % Equation 7.26

 if bat_type=='NC'
    E=open_circuit_voltage_NC(DoD(C-1),NoCells);
 elseif bat_type=='LA'
    E=open_circuit_voltage_LA(DoD(C-1),NoCells);
 else
     error('Invalid battery type');
 end;

if Pbat > 0 % Use Equ. 2.20
    I = (E - ((E*E) - (4*Rin*Pbat))^0.5)/(2*Rin);
    CR(C) = CR(C-1) +((I^k)/3600);    %Equation 2.18
 elseif Pbat==0
      I=0;
 elseif Pbat <0
   % Regenerative braking. Use Equ. 2.22, and
   % double the internal resistance.
   Pbat = - 1 * Pbat;
   I = (-E + (E*E + (4*2*Rin*Pbat))^0.5)/(2*2*Rin);
   CR(C) = CR(C-1) - (I/3600);    %Equation 2.23
end;

DoD(C) = CR(C)/PeuCap;    %Equation 2.19
if DoD(C)>1
   DoD(C) =1;
end
```

```
% Since we are taking one second time intervals,
% the distance traveled in metres is the same
% as the velocity. Divide by 1000 for km.
  D(C) = D(C-1) + (V(C)/1000);
  XDATA(C)=C;        % See Section 7.4.4 for the use
  YDATA(C)=eff_mot;  % of these two arrays.
end;
% Now return to calling program.
```

Appendix 4: Range Simulation of the GM EV1 Electric Car

In Section 7.4.2.3 the simulation of this important vehicle was discussed. Figure 7.15 gives an example output from a range simulation program. The MATLAB® script file for this is shown below. Notice that it calls several of the MATLAB® files we have already described. However, it should be noted how this program sets up, and often gives values to, the variables used by the program one_cycle described in the preceding section.

```
% Simulation of the GM EV1 running the SFUDS
% driving cycle. This simulation is for range
% measurement. The run continues until the
% battery depth of discharge > 90%

sfuds;    % Get the velocity values, they are in
          % an array V.
N=length(V); % Find out how many readings
%Divide all velocities by 3.6, to convert to m/sec
V=V./3.6;

% First we set up the vehicle data.
mass = 1540 ;   % Vehicle mass+ two 70 kg passengers.
area = 1.8;     % Frontal area in square metres
Cd = 0.19;      % Drag coefficient
Gratio = 37;    % Gearing ratio, = G/r
G_eff = 0.95;   % Transmission efficiency
Regen_ratio = 0.5; % This sets the proportion of the
            % braking that is done regeneratively
            % using the motor.
bat_type='LA';  % Lead acid battery
NoCells=156;    % 26 of 6 cell (12 Volt) batteries.
Capacity=60;    % 60 Ah batteries. This is
            % assumed to be the 10 hour rate capacity
k=1.12;     % Peukert coefficient, typical for good lead acid
Pac=250;    % Average power of accessories.

% These are the constants for the motor efficiency
% equation, 7.23
kc=0.3;       % For copper losses
ki=0.01;      % For iron losses
kw=0.000005;  % For windage losses
ConL=600;     % For constant electronics losses
```

```matlab
% Some constants which are calculated.
Frr=0.0048 * mass * 9.8;    % Equation 7.1
Rin= (0.022/Capacity)*NoCells;    % Int. res, Equ. 2.2
Rin = Rin + 0.05;  % Add a little to make allowance for
                   % connecting leads.
PeuCap= ((Capacity/10)^k)*10;  % See equation 2.12
% Set up arrays for storing data for battery,
% and distance traveled. All set to zero at start.
% These first arrays are for storing the values at
% the end of each cycle.
% We shall assume that no more than 100 of any cycle is
% completed. (If there are, an error message will be
% displayed, and we can adjust this number.)
DoD_end = zeros(1,100);
CR_end = zeros(1,100);
D_end = zeros(1,100);

% We now need similar arrays for use within each cycle.
DoD=zeros(1,N);   % Depth of discharge, as in Chap. 2
CR=zeros(1,N);    % Charge removed from battery, Peukert
                  % corrected, as in Chap 2.
D=zeros(1,N);     % Record of distance traveled in km.

CY=1;
% CY controls the outer loop, and counts the number
% of cycles completed. We want to keep cycling till the
% battery is flat. This we define as being more than
% 90% discharged. That is, DoD_end > 0.9
% We also use the variable XX to monitor the discharge,
% and to stop the loop going too far.
DD=0;   % Initially zero.

while DD < 0.9
%Beginning of a cycle.************
% Call the script file that performs one
% complete cycle.

one_cycle;

% One complete cycle done.
% Now update the end of cycle values.
DoD_end(CY) = DoD(N);
CR_end(CY) = CR(N);
D_end(CY) = D(N);

% Now reset the values of these "inner" arrays
% ready for the next cycle. They should start
% where they left off.

DoD(1)=DoD(N); CR(1)=CR(N);D(1)=D(N);
DD=DoD_end(CY)   % Update state of discharge
%END OF ONE CYCLE **************
CY = CY +1;
end;
```

```
plot(D_end,DoD_end,'k+');
ylabel('Depth of discharge');
xlabel('Distance traveled/km');
```

The plot lines at the end of the program produce a graph such as in Figure 7.15. This graph has two sets of values. This is achieved by running the program above a second time, using the MATLAB® hold on command. The second running was with much a higher value (800) for the average accessory power P_{ac}, and a slightly higher value (1.16) value for the Peukert Coefficient.

Appendix 5: Electric Scooter Range Modelling

By way of another example, the MATLAB® script file below is for the range modelling of an electric scooter. The program is very similar, except that almost all the variables are different, and a different driving cycle is used. This shows how easy it is to change the system variables to simulate a different vehicle.

```
% Simulation of the electric scooter running
% the ECE-47 driving cycle. This simulation is
% for range measurement. The run continues until
% the battery depth of discharge > 90%

ECE_47; % Get the velocity values, they are in
% an array V, and in m/sec.

N=length(V); % Find out how many readings

% First we set up the vehicle data.
mass = 185 ;      % Scooter + one 70 kg passenger.
area = 0.6;       % Frontal area in square metres
Cd = 0.75;        % Drag coefficient
Gratio = 2/0.21;  % Gearing ratio, = G/r
G_eff = 0.97;     % Transmission efficiency
Regen_ratio = 0.5; %This sets the proportion of the
                  % braking that is done regeneratively
                  % using the motor.
bat_type='NC';    % NiCAD battery.
NoCells=15;       % 3 of 5 cell (6 Volt) batteries.
Capacity=100;     % 100 Ah batteries. This is
                  % assumed to be the 5 hour rate capacity
k=1.05;           % Peukert coefficient, typical for NiCad.
Pac=50;           % Average power of accessories.

kc=1.5;    % For copper losses
ki=0.1;    % For iron losses
kw=0.00001; % For windage losses
ConL=20;   % For constant motor losses

% Some constants which are calculated.
Frr=0.007 * mass * 9.8;  % Equation 7.1
```

Appendices: MATLAB® Examples

```
Rin = (0.06/Capacity)*NoCells; % Int. resistance, Equ. 2.2
Rin = Rin + 0.004; %Increase int. resistance to allow for
                  % the connecting cables.
PeuCap = ((Capacity/5)^k)*5 % See equation 2.12
% Set up arrays for storing data for battery,
% and distance traveled. All set to zero at start.
% These first arrays are for storing the values at
% the end of each cycle.
% We shall assume that no more than 100 of any cycle is
% completed. (If there are, an error message will be
% displayed, and we can adjust this number.)
DoD_end = zeros(1,100);
CR_end = zeros(1,100);
D_end = zeros(1,100);

% We now need similar arrays for use within each cycle.
DoD=zeros(1,N);   % Depth of discharge, as in Chap. 2
CR=zeros(1,N);    % Charge removed from battery, Peukert
                  % corrected, as in Chap 2.
D=zeros(1,N);     % Record of distance traveled in km.
XDATA=zeros(1,N);
YDATA=zeros(1,N);
CY=1;
% CY controls the outer loop, and counts the number
% of cycles completed. We want to keep cycling till the
% battery is flat. This we define as being more than
% 90% discharged. That is, DoD_end > 0.9
% We also use the variable XX to monitor the discharge,
% and to stop the loop going too far.
DD=0;   % Initially zero.

while DD < 0.9
%Beginning of a cycle.************

one_cycle;
% **********
% Now update the end of cycle values.
DoD_end(CY) = DoD(N);
CR_end(CY) = CR(N);
D_end(CY) = D(N);

% Now reset the values of these "inner" arrays
% ready for the next cycle. They should start
% where they left off.

DoD(1)=DoD(N); CR(1)=CR(N);D(1)=D(N);
DD=DoD_end(CY)   % Update state of discharge
%END OF ONE CYCLE **************
CY = CY +1;
end;
plot(XDATA,YDATA,'k+');
```

Notice that the last line plots data collected during one cycle, as explained in Section 7.4.4. Graphs such as Figure 7.16 and 7.17 were produced in this way.

If we wish to find the range to exactly 80% discharged, then the `while DD < 0.9;` line is changed to `while DD < 0.8;` the following line is added to the end of the program in place of the plot command.

```
Range = D(N)*0.8/DoD(N)
```

The lack of the semicolon at the end of the line means that the result of the calculation will be printed, without the need for any further command. Results such as those in Table 7.3 were obtained this way.

Appendix 6: Fuel Cell Range Simulation

In Section 7.4.5 the question of the simulation of vehicle range simulation was discussed in relation to fuel cells. In the case of systems with fuel reformers it was pointed out that such simulations are highly complex. However, if the hydrogen fuel is stored as hydrogen, and we assume (not unreasonably) that the fuel cell has more-or-less constant efficiency, then the simulation is reasonably simple. An example, which is explained in Section 7.4.5 is given below.

```
% Simulation of a GM EV1 modified to incorporate
% a fuel cell instead of the batteries, as outlined
% in section 7.4.5.
% All references to batteries can be removed!

sfuds; % Get the velocity values, they are in
       % an array V.
N=length(V); % Find out how many readings
%Divide all velocities by 3.6, to convert to m/sec
V=V./3.6;

% First we set up the vehicle data.
mass = 1206 ; % Vehicle mass + two 70 kg passengers.
              % See section 7.4.5
area = 1.8;    % Frontal area in square metres
Cd = 0.19;     % Drag coefficient
Gratio = 37;   % Gearing ratio, = G/r
Pac=2000;      % Average power of accessories. Much larger,
     % as the fuel cell needs a fairly complex controller.
kc=0.3;    % For copper losses
ki=0.01;   % For iron losses
kw=0.000005;  % For windage losses
ConL=600;  % For constant electronics losses

% Some constants which are calculated.
Frr=0.0048 * mass * 9.8;   % Equation 7.1
 % Set up arrays for storing data.
 % Rather simpler, as hydrogen mass left
 % and distance traveled is all that is needed.
% This first array is for storing the values at
% the end of each cycle.
```

```
% We need many more cycles now, as the range
% will be longer. We will allow for 800.

D_end=zeros(1,800);
H2mass_end = zeros(1,800);
H2mass_end(1) =8.5; % See text. 8.5 kg at start.

% We now need a similar array for use within each cycle.
D=zeros(1,N);
H2mass=zeros(1,N);   % Depth of discharge, as in Chap. 2
H2mass(1)=8.5;

CY=1;
% CY defines is the outer loop, and counts the number
% of cycles completed. We want to keep cycling till the
% the mass of hydrogen falls to 1.7 kg, as explained in
% Section 7.4.5.
% We use the variable MH to monitor the discharge,
% and to stop the loop going too far.
MH=8.5;  % Initially full, 8.5 kg

while MH > 1.7
   %Beginning of a cycle.************
for C=2:N
   accel=V(C) - V(C-1);
   Fad = 0.5 * 1.25 * area * Cd * V(C)^2; % Equ. 7.2
   Fhc = 0;                % Eq. 7.3, assume flat
   Fla = 1.01 * mass * accel;
      % The mass is increased modestly to compensate for
   % the fact that we have excluded the moment of inertia
   Pte = (Frr + Fad + Fhc + Fla)*V(C); %Equ 7.9 & 7.23
   omega = Gratio * V(C);
   if omega == 0
       Pte=0;
       Pmot=0;
       Torque=0;

   elseif omega > 0
       Torque=Pte/omega; % Basic equation, P = T * ω
       if Torque>=0
            % Now equation 7.23
          eff_mot=(Torque*omega)/((Torque*omega) +
          ((Torque^2)*kc)+ (omega*ki)+((omega^3)*kw)+ConL);
       elseif Torque<0
          eff_mot =(-Torque*omega)/((-Torque*omega) +
          ((Torque^2)*kc) +( omega*ki)+((omega^3)*kw)+ConL);
       end;
       if Pto > = 0
          Pmot = Pte/(0.9 * eff_mot);   % Equ 7.23
       elseif Pte < 0
          % No regenerative braking
          Pmot = 0;
       end;
   end;
```

```
        Pfc = Pmot + Pac;

        H2used = 2.1E-8 * Pfc; % Equation 7.29
        H2mass(C) = H2mass(C-1) - H2used; %Equation 7.29 gives
        % the rate of use of hydrogen in kg per second,
        % so it is the same as the H2 used in one second.

        % Since we are taking one second time intervals,
        % the distance traveled in metres is the same
        % as the velocity. Divide by 1000 for km.
        D(C) = D(C-1) + (V(C)/1000);
    end;
    % Now update the end of cycle values.
    H2mass_end(CY) = H2mass(N);
    D_end(CY) = D(N);

    % Now reset the values of these "inner" arrays
    % ready for the next cycle. They should start
    % where they left off.

    H2mass(1)=H2mass(N); D(1)=D(N);
    MH=H2mass_end(CY)   % Update state of discharge
    %END OF ONE CYCLE ***************
    CY = CY +1;
end;
% Print the range.
Range = D(N)
```

Appendix 7: Motor Efficiency Plots

In Chapter 6 the question of efficiency plots of electric motors was addressed. An example was given in Figure 6.7. It is very useful to be able to print this sort of diagram, to see the operating range of electric motors, and where they operate most efficiently. Furthermore, this can be done very effectively and quickly with MATLAB®. The script file is given below.

```
% A program for plotting efficiency contours for
% electric motors.
% The x axis corresponds to motor speed (w),
% and the y axis to torque T.
% First, set up arrays for range.
x=linspace(1,180);% speed, N.B. rad/sec NOT rpm
y=linspace(1,40); % 0 to 40 N.m
% Allocate motor loss constants.
kc=1.5;     % For copper losses
ki=0.1;     % For iron losses
kw=0.00001; % For windage losses
ConL=20;    % For constant motor losses
% Now make mesh
[X,Y]=meshgrid(x,y);

Output_power=(X.*Y); % Torque x speed = power
B=(Y.^2)*kc;  % Copper losses
```

```
C=X*ki;         % Iron losses
D=(X.^3)*kw;    % Windage losses
Input_power = Output_power + B + C + D + ConL;
Z = Output_power./Input_power;
% We now set the efficiencies for which a contour
% will be plotted.
V=[0.5,0.6,0.7,0.75,0.8,0.85,0.88];
box off
grid off

contour(X,Y,Z,V);
xlabel('Speed/rad.s^-^1'), ylabel('Torque/N.m');

hold on
% Now plot a contour of the power output
% The array Output_Power has
% already been calculated. We draw contours at
% 3 and 5 kW.
V=[3000,5000];
contour(X,Y,Output_power,V);
```

This program was used to give the graph shown as Figure 6.7. In Figure A.1 we shown the result obtained for a higher power, higher speed motor, without brushes. All that has happened is that the motor loss constants have been changed, and the ranges of values for torque and angular speed have been increased as follows:

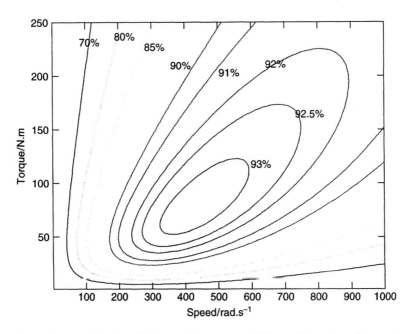

Figure A.1 A plot showing the efficiency of a motor at different Torque/speed operating points. It shows the circular contours characteristic of the brushless DC motor

```
x=linspace(1,1000);  % N.B. rad/sec, not rpm
y=linspace(1,250);
% Allocate motor loss constants.
kc=0.2;     % For copper losses
ki=0.008;   % For iron losses
kw=0.00001; % For windage losses
ConL=400;   % For constant motor losses
```

Also, the values of efficiency were changed, and the last lines that plot constant power contours were removed.

Index

4 quadrant controller, 8, 241

Acid electrolyte fuel cell, 84
Aerodynamics, 185, 213, 217, 268
Air conditioning, 239
Alkaline fuel cells, 86
Ammonia, 135
Amphour capacity
    effect of higher currents on, 57
    modeling, 57
    term explained, 25
Apollo spacecraft, 86
Armature, 142
Autothermal reforming, 116, 118, 128

Balance of plant, 107
Batteries
    charge equalisation, 49, 50
    different types compared, 52, 67
    equivalent circuit, 24, 55
    Modeling, 54
Battery charging, 35, 48, 244
Battery electric vehicles
    applications, 5, 8, 262, 263
    effect of mass on range, 224
    emissions from, 250, 251
    examples, 8, 189, 193, 261, 265
    performance modeling, 189, 193, 279
    range modeling, 201, 218, 224, 284, 286
    simulation, 207
Battery life, 49
Bicycles, 261
Bipolar plates, 96
Blowers, 107
Body design, 226, 228
Brushless DC motor, 167, 275
Buses, 16, 19, 83, 272

C notation, 25
California air resources board, 12, 48, 259, 268
Capacitors, 19, 74
Carbon monoxide, 246
    removal, 117
Carbon nanofibres, 126
Carnot limit, 89, 92
Catalysts, 87, 116, 136
Charge equalisation, 49, 50, 75
Charging, 35, 48, 244
Charging efficiency, 28, 50
Chassis design, 226, 228
Chassis materials, 230
Chopper circuits *See* DC/DC converters, 157
Coefficient of rolling resistance, 184, 218
Comfort, 231, 237, 243
Commutator, 142
Compressors, 107
Controls, 240
Cooling, 238
Copper losses, 149
Crash resistance, 228

DC/DC converters, 108, 155
    efficiency of, 159, 161
    step-down, 157
Digital signal processors, 171
Direct methanol fuel cells, 85
Drag coefficient, 185, 214
Driving cycles, 196, 280
    10–15 Mode, 196
    ECE-15, 196
    ECE-47, 199, 206, 281
    EUDC, 196
    FHDS, 196
    FUDS, 196
    MATLAB, 280

Driving cycles (*continued*)
  SAE J227, 198
  SFUDS, 196, 205, 211
Driving schedules *See* Driving cycles, 196
Dynamic braking, 153, 275

Efficiency
  DC/DC converters, 158, 161
  limit for fuel cells, 92
  motors, 149, 175, 202, 290
  of fuel cells defined, 91
Electric scooters, 189, 200, 206, 286
Electronic switches, 155, 156
Emission
  from different vehicle types compared, 251
Energy density
  Batteries and fuel compared, 3
  term explained, 3
Enthalpy, 90, 91
Equivalent circuit
  batteries, 24, 55
Ethanol, 129, 245, 258
Exergy, 90

Faraday, unit of charge, 90
Flywheels, 18, 54, 72
Ford, 266, 276
Fuel cell powered vehicles
  examples, 17, 83, 272
Fuel cell vehicles
  buses, 16
  cars, 16
  emissions from, 249, 258, 259
  examples, 15, 17, 83, 272
  main problems, 81
  range modeling, 208, 288
Fuel cells
  basic chemistry, 84
  cooling, 105, 108, 273
  different types (table), 84
  efficiency, 91
  efficiency defined, 92
  efficiency/voltage relation, 92
  electrodes, 87
  leaks, 101
  Nernst equation, 96
  osmotic drag, 104
  pressure, 96
  reversible voltage, 92
  temperature, 87, 92
  thermodynamics, 91
  voltage/current relation, 94
  water management, 101, 104

Gasoline
  use with fuel cells, 118
Geothermal energy, 257
Gibbs free energy
  changes with temperature, 91
  explained, 90
GM EV1, 193, 205, 211, 215, 239, 267, 279, 284
GM Hy-wire, 107, 226, 233, 241
Greenhouse effect, 247

Harmonics, 163
Heat pumps, 239
Heating, 237, 238
High pressure hydrogen storage, 120, 122
Hill climbing, 185, 224
Hindenburg, 120
History, 1
Honda Insight, 53, 179, 217, 232, 269
Hybrid electric vehicles
  battery charge equalisation, 50
  battery selection, 53
  electrical machines, 179
  emissions from, 250, 251, 259
  examples, 13, 269, 271
  grid connected, 259
  parallel, 10, 180, 270
  series, 10
  supply rails, 79
  term explained, 9
  with capacitors, 19, 77
Hydroelectricity, 255
Hydrogen
  as energy vector, 124
  from gasoline, 118
  from reformed methanol, 115, 117
  made by steam reforming, 114
  physical properties, 120
  safety, 120, 122–124
  storage as a compressed gas, 120, 275
  storage as a cryogenic liquid, 122
  storage in alkali metal hydrides, 130
  storage in chemicals, 127
  storage in metal hydrides, 124
  storage methods compared, 138
Hydrogen fueled ICE vehicle, 249, 259

IGBTs, 157
Induction motor, 173
Inductive power transfer, 78
Internal resistance, 24, 30, 38
Inverters
  3-phase, 165
Iron losses, 149

Index

Kamm effect, 217

Lead acid batteries
   basic chemistry, 30, 32
   internal resistance, 30
   limited life, 34
   main features, 31
   modeling, 56
   sealed types, 32
Liquid hydrogen, 122
Lithium batteries
   basic chemistry, 45
   main features, 45
Low speed vehicles, 263

Marine current energy, 257
Materials selection, 230, 232
Metal air batteries
   aluminium/air, 46
   zinc/air, 47
Metal hydride storage of hydrogen, 124
Methanation of carbon monoxide, 117
Methane, 116, 120
Methanol, 250, 259
   as hydrogen carrier, 115, 130, 134
Methanol fuel cell, 85
Mobility aids, 263
Molten carbonate fuel cell, 86
MOSFETs, 156
Motors
   BLDC, 275
   brushed DC, 141
   brushless DC, 167
   copper losses, 149
   efficiency, 149, 175, 202, 290
   fuel cells, used with, 108
   induction, 173
   integral with wheel, 180, 221, 223
   iron losses, 149
   mass of, 177
   power/size relation, 151
   self-synchronous, 167
   specific power, 177
   switched reluctance, 169
   torque/speed characteristics, 143

Nafion, 102
Nickel cadmium batteries
   basic chemistry, 36
   charging, 37
   internal resistance, 38
   main features, 37
   modeling, 56
Nickel metal hydride batteries
   applications, 41
   basic chemistry, 39
   main features, 39
Nuclear energy, 257

Orbiter spacecraft, 86
Osmotic drag, 104

Partial oxidation reformers, 116, 118
PEM fuel cells
   electrode reactions, 84
   electrolyte of, 101
   introduced, 85
   reformed fuels, use with, 115
Perfluorosulphonic acid, 102
Performance modeling, 188
Peugeot, 189, 200, 266
Peukert Coefficient, 57, 64, 203
Phosphoric acid fuel cells, 86
Pollution, 245, 248, 251, 259
Power steering, 243
Propane, 120
Proton exchange membrane, 84, 101
PTFE, 102

Rear view mirrors, 243
Regenerative braking, 9, 153, 206, 225, 270
Regulators, 155, 157, 159
Rolling resistance, 184, 218

Selective oxidation reactor, 117
Self discharge of batteries, 32
Shift reactors See Water gas shift reaction, 117
Shuttle spacecraft See Orbiter spacecraft, 86
Sodium borohydride
   as hydrogen carrier, 132
   cost, 135
Sodium metal chloride batteries See Zebra
   batteries, 42
Sodium sulphur batteries
   basic chemistry, 41
   main features, 42
Solar energy, 18, 69, 254
Solid oxide fuel cells, 86
Specific energy
   relation to specific power, 28
   term explained, 27
Stability, 227
Stack, 96
Steam reforming, 114, 118
Sulphonation, 102
Super-capacitors See Capacitors, 19
Supply rails, 18, 77

Suspension, 231
Switched reluctance motors, 169

Thyristors, 157
Tidal energy, 255
Total energy use, 254
Toyota Prius, 13, 41, 53, 271
Tractive effort, 187
Transmission, 221
Types of fuel cell (table), 85
Tyre choice, 243

Ultra-capacitors *See* Capacitors, 19

Water gas shift reaction, 114, 117
Watthour
   term explained, 26
Well-to-wheel analysis, 248, 251
Wind energy, 71, 255
Windage losses, 150

Zebra batteries
   basic chemistry, 42
   main features, 43
   operating temperature, 43
Zinc air batteries, 16

Printed in the United States
96286LV00003B/157/A